William Barton Rogers and the Idea of MIT

*W*ILLIAM *B*ARTON *R*OGERS
and the Idea of MIT

A. J. ANGULO

The Johns Hopkins University Press
Baltimore

© 2009 The Johns Hopkins University Press
All rights reserved. Published 2009
Printed in the United States of America on acid-free paper
2 4 6 8 9 7 5 3 1

The Johns Hopkins University Press
2715 North Charles Street
Baltimore, Maryland 21218-4363
www.press.jhu.edu

Library of Congress Cataloging-in-Publication Data
Angulo, A. J.
William Barton Rogers and the idea of MIT / A. J. Angulo.
p. cm.
Includes bibliographical references and index.
ISBN-13: 978-0-8018-9033-8 (hbk. : alk. paper)
ISBN-10: 0-8018-9033-0 (hbk. : alk. paper)
1. Rogers, William Barton, 1804–1882. 2. Massachusetts Institute
of Technology—History. 3. Massachusetts Institute of Technology—
Presidents—Biography. 4. College presidents—Massachusetts—Biography.
5. Science—Study and teaching (Higher)—United States—History.
6. Engineering—Study and teaching—United States—History. I. Title.
T171.M495R73 2008
378.0092—dc22   2008013749

A catalog record for this book is available from the British Library.

The MIT seal that appears on page 157 is used with permission.

*Special discounts are available for bulk purchases of this book. For more information, please
contact Special Sales at 410-516-6936 or specialsales@press.jhu.edu.*

The Johns Hopkins University Press uses environmentally friendly book materials,
including recycled text paper that is composed of at least 30 percent post-consumer
waste, whenever possible. All of our book papers are acid-free, and our jackets and
covers are printed on paper with recycled content.

To Acorn and Asa

CONTENTS

The commencement speech didn't last long, but it remains one of the most memorable in academic history. If he spoke slowly, it might have lasted three to four minutes. William Barton Rogers, conceptual founder of the Massachusetts Institute of Technology, was not exactly known for his brevity, but then again few could have predicted how the day was to turn out.

He had come to talk about the origins of MIT at its commencement ceremony of 1882. It was a typical audience. There were soon-to-be graduates, most of them anxious to get their degrees. Some of them squirmed in their seats at the thought of presenting an abstract of their senior theses, as required for graduation. There were supporting family members who came to watch their sons and daughters present their research and receive their diplomas. The ceremony attracted members of the community as well. Many were curious to know more about this emerging institution located then in Boston's Back Bay. Joining this crowd were friends and admirers of the speaker who had come just to hear him talk. Rogers knew he wanted to focus on the foundation of MIT and the resistance he had initially faced from leaders in traditional higher education. But as with most of his speeches, he left plenty of room for improvisation. After being introduced by the Institute's president, Francis Amasa Walker, Rogers stood before the gathering in Huntington Hall and began to speak with pride about what MIT had become. He shared with the audience the early struggles, the mixed reception it had gotten from educational leaders, and the founding mission of offering a comprehensive program of scientific and engineering studies. "Formerly a wide separation existed between theory and practice," he reminisced. "Now in every fabric that is made, in every structure that is reared, they are closely united into one interlocking system—the practical is based upon the scientific, and the scientific is solidly built upon the practical." Partway into the speech he paused, briefly glanced at his notes, and then foundered at the knees. By the time he fell to the platform, Rogers was dead.[1]

It could hardly be more appropriate that his life came to a close in this way, not necessarily for the drama of it but more in that it captured the seriousness of his life-long passion for science, enthusiasm for higher learning, and relentless work ethic. The title of a well-known Isaac Newton biography, *Never at Rest,* could just as easily work for Rogers, who had a long and productive career as a scientist and educational reformer. His commitment to science drove him, and many others of his generation, to do things we could rarely imagine doing today. He conducted Virginia's first state geological survey by foot, horse, and buggy from 1835 to 1842. For those seven years Rogers climbed mountains and cliffs, waded through swamps, and endured many hardships and one tragic death among his small team of assistants, all the while collecting samples and constructing a comprehensive geological map that retained its scientific value for decades. The survey was one of many projects he undertook during the first half of his career in Virginia and second half in Massachusetts. To the very end he continued to prepare papers and presentations and showed few signs of slowing down.[2]

The same is true of his educational reform efforts. As early as his first full-time teaching appointment in Maryland, Rogers experimented with ways of communicating scientific ideas to his students. Traditional, lecture-based modes of science instruction bothered him like pebbles in his shoes. When he went to teach at the College of William and Mary, he tried out some alternative methods there and later at the University of Virginia. What he learned, or at least believed, as a result was that a new kind of institution was needed to educate American scientists and engineers. The classical colleges of the first half of the nineteenth century failed to satisfy his desire for getting scientific breadth and depth, theory and practice into the curriculum. For decades he turned over in his mind the ideas that led to the establishment of MIT. In the process he wrote proposals for politicians, philanthropists, and educational leaders to consider the kind of educational reform he believed was necessary for the advancement of American science. After several attempts, "sometimes met not only with repulse but with ridicule," as Rogers described in his commencement speech, he eventually found a home for his ideas in Massachusetts. Boston politicians gave into the concept of an institute of technology in the 1860s.[3]

He wrote convincingly on scientific matters and for the cause of educational reform, but perhaps his greatest asset was his public speaking ability. That's why his final moments seem so appropriate. It only makes sense that his last words would be about these passions. Rogers had a way of persuading others through conversation, lectures, and debates. In the legislative halls of Richmond he knew how to sell the idea of a geological survey to Virginians; in Boston his way with words forged connections between powerful interests and the idea of progress through an institute of technology. Victorian society generously rewarded those with golden tongues. His was at least silver.

Rogers's most enduring reward, the Massachusetts Institute of Technology, stands today in Cambridge, just across the Charles River from where it first began. The classical dome-like structures can be seen from Boston and can't be missed when traveling across the Harvard Bridge. Its physical presence is unmistakable, with over 150 acres along the Charles. Its presence in the higher education landscape is equally striking. While it has long been a relatively small campus in terms of enrollment, it's had a significant impact through its research and outreach educational programs. At the start of the twenty-first century MIT received approximately a half-billion dollars for sponsored research. These dollars helped support projects that tackled basic and practical problems in areas such as energy, defense, health, and industry. Likewise, at the start of the century the Institute took notable steps toward having an equally visible impact on education and outreach. Through the OpenCourseWare initiative MIT has made course content and materials available online and free of charge. Scholars around the world have commented on the resources provided by this initiative and the long-term impact it will likely have on the advancement of science curriculum and pedagogy.

As significant as these achievements are in their own right, there's nothing all that new about the basic ideals undergirding them. We can see the ideals over 150 years ago, fueling Rogers's lifelong passion for the advancement and diffusion of scientific knowledge that led to the founding of MIT in the first place. Rogers followed closely European advances in science; he read widely to keep current with the geological and natural philosophical works of the French, German, and British. His own papers incorporated their insights and looked for ways to build on the latest developments. Not surprisingly, he made sure student and faculty research stood at the center of the Institute's mission. At the same time, he had an egalitarian, some might say American, drive to bring these advances to the public. A generation earlier Thomas Jefferson had unsuccessfully attempted to give expression to this sentiment with his characteristically democratic *Bill for the More General Diffusion of Knowledge*. Rogers shared with Jefferson a similar Enlightenment belief that direct improvements in the lives of all would occur through the diffusion of knowledge. To this end Rogers's plan for the institute included a push for faculty to offer free lecture hall classes for the general public in such areas as mathematics, chemistry, and physics. His desire for an institution that would balance advancement and diffusion through research, teaching, and service seems largely out of place in an era when even the best colleges in the nation had, as one nineteenth-century scientist put it, "more the character of a high school than a University."[4]

With the establishment of the Institute, Rogers and his circle of reformers helped usher in a new era in higher education history. Historians have paid much attention to the English and Germanic influences on American colleges and universities and they have long described how the British classical system was adopted nearly whole-sale at the first undergraduate colonial institutions: Harvard, William and Mary, and Yale. Some accounts continue the story of British influence from the seventeenth century until the arrival in the nineteenth century of German research and *Wissenschaft*. The institution most strongly associated with this change is the Johns Hopkins University, which began offering graduate-level education in 1876. Less attention, however, has been paid to the French and their polytechnic systems of science education. Rogers was profoundly influenced by French scientific and engineering education, and the influence figures prominently in the conceptual organization of the institute. MIT offers a compelling case with which to rethink our British-German paradigm.

This biography offers an account of the wide-ranging scientific and educational values Rogers sustained throughout his life, values that made the institute what it was in its early years and continue to guide it to this day. To bring together these diverse strands of his life, this study presents Rogers in a largely thematic form, each theme placing Rogers within the social and intellectual context of his era. His experiences in Virginia, activities in science, and vision for higher learning each receive attention separately, although at times the points converge. His life spanned nearly the entire nineteenth century, including such social transformations as the onset of industrialization, the spawning of reform movements, the hardening of southern civilization, the Civil War, the reconstruction of the South, the maturation of industrialism, and the start of Progressivism. Revolutions in American intellectual life were equally dramatic. Rogers lived through a fundamental shift experienced by virtually all scientists of his generation, a shift marked by the decline of the generalist and the rise of the specialist. At the start of Rogers's career in science his teaching and research reflected a generalist approach to science, as demanded by colleges and universities. Science professors of the first decades of the century could be found teaching everything from algebra to zoology. By the end of his career, however, Rogers pointed toward a model of higher learning that required its faculty to specialize. Coupled with this shift was the publication of Charles Darwin's *On the Origin of Species,* a work that appeared at the peak of Rogers's professional career. After the book's publication in 1859, virtually no field in science, whether general or specialized, escaped the implications of natural selection.

Owing to the diversity of Rogers's interests, readers may find some themes more compelling than others. Historians of education are likely to focus on the dilemma of southern higher learning or the origins of MIT. More important to historians of science is the treatment of geology and natural philosophy, the professionalization of science, and the conflicts Rogers had with the Lazzaroni. Still other scholars with interests in higher learning and instruction may gravitate toward the discussion of Rogers's ideas about reform or the diffusion of innovations, especially with regard to the laboratory. Readers interested in the interrelationships between these themes, however, should allow themselves to wander with the subject to the shores, mountains, and lecture halls of Virginia, to the bustle of Boston and the Back Bay, and on occasion to Europe.

Biographies, of course, are not the best place to settle long-standing squabbles among historians, and this study makes no pretenses to the contrary. One life can illuminate a case study, but it hardly makes a conclusive argument. The pages that follow will not provide definitive answers to questions about slavery's impact on science and higher education. Rather, Rogers's experiences in the antebellum South trumpet a call for further research about others in similar positions who shared his views. Likewise, these pages do not attempt a complete history of MIT's origins, much less a history of technological institutes. One scholar's decades-old protest that "the history of technical education in America remains to be written" deserves repeating.[5]

What this study does contain is an analysis of the way Rogers went about the business of science and higher education. It tracks a life that began in Pennsylvania, matured in Virginia, and culminated in Massachusetts. The analysis presented here portrays a life governed by a conviction about the value of both theory and practice, rather than an exclusive interest in one or the other, as many of his generation tended to do. The conviction is described in this study as the ideal of the useful arts.

This biography relies on many general histories to develop the theme of the useful arts in Rogers's life. The reader should take seriously the notes and bibliography provided, for this study could not have been possible without the insights derived from the historians listed there. As for works directly related to Rogers, very little exists. An uneven assortment of articles and book chapters have recited a chronology of milestones in his life. But with the exception of Emma Savage's Victorian-style *Life and Letters of William Barton Rogers,* no extended inquiry into his life has been published. This study owes much to Savage's two-volume compendium of her husband's letters and to her patient deciphering of Rogers's notorious scrawl. Like the scores of other *Life and Letters* compilations, however, Savage's could not escape the distortions and omissions expected of such volumes. The genre, committed to casting the best light on their subjects, reflects the didactic tendencies of Victorian era life writing. Where possible, this study relies instead on the rich collection of his papers located at the MIT archives as well as the less plentiful repositories in Virginia and elsewhere.[6]

⇝⇜

This study would not have been possible without the advice and encouragement of many individuals. Thomas G. Dyer introduced me to the fields of history of higher education and southern history. His work in both areas continues to serve as models for my own work. Derrick P. Alridge gave me an introduction to history of education broadly conceived. I finally met W. E. B. DuBois through him and, in the process, learned the significance of biography in relation to historiography. Julie Reuben, my doctoral advisor, read this book in its first incarnation, and I greatly benefited from her broad knowledge of science in nineteenth-century America. Her work on the formation of the modern university gave me a vital starting point for considering the origins of the Massachusetts Institute of Technology. Catherine Z. Elgin unpacked the arguments in the book and offered a refreshingly analytical perspective on their strengths and weaknesses. She has had a lasting impact on my approach to thinking, writing, teaching, and ways of understanding. Thomas A. Underwood read and commented on the book at two critical points in its development. He generously gave of his time and intellectual energy, and I am grateful to have had his feedback. Thomas F. Glick read an earlier version of the work, and his suggestions led to many fruitful explorations of the literature on evolution and nineteenth-century science.

Many others have taken an interest in this book, commenting on some or all of its elements and themes. Edward Jones-Imhotep, Patricia A. Graham, David Tyack, Robert Brain, and William Gienapp read select chapters and shared their valuable reflections on William Barton Rogers, MIT, and nineteenth-century America. I want to thank Roger Geiger of *Perspectives on the History of Higher Education* and the anonymous reviewers of the *History of Education Quarterly* for their comments on research drawn from this study. To the editors of both publications, I give thanks for permission to reprint portions of my work that appeared in their journals: Parts of chapter 2 appeared in "William Barton Rogers and the Southern Sieve: Revisiting Science, Slavery, and Higher Learning in the Old South," *History of Education Quarterly* 45 (March 2005): 18–37. Parts of chapter 8 appeared in "The Initial Reception of MIT, 1861–1882," *Perspectives on the History of Higher Education* (formerly *History of Higher Education Annual*) 26 (2007): 1–28. While I owe much to this circle of advisors, researchers, and friends, the limitations and weaknesses that remain in this work are mine alone.

I am also thankful for the institutional support that has aided my work in essential ways. The staff at the following archives and sites were particularly helpful and supportive of my research efforts: Liz Andrews and Nora Murphy of the Massachusetts Institute of Technology Archives; Gina Woodward and Margaret C. Cook of the Special Collections Division of the Swem Library at the College of William and Mary; Regina Rush at the University of Virginia Special Collections Department; and Daniel Barbiero at the National Academy of Science Archives; Lydia Carey and the rest if the team at the NotSo Hostel in Charleston, South Carolina. The friendly assistance I received at the Harvard University Archives, Houghton Library, Special Collections at the Museum of Comparative Zoology, and Massachusetts Historical Society facilitated the research for this study. At Harvard the Roy E. Larsen Fellowship, the John E. Thayer Scholarship, and the C. V. Starr Scholarship provided me with sustained periods for reflection. At my present institution, Winthrop University, I have benefited from colleagues who have accommodated my research needs and teaching schedule requests. Their support has contributed to the completion of this study.

William Barton Rogers and the Idea of MIT

CHAPTER ONE

An Uncertain Future

ALMOST AS SOON AS he'd finished a course of studies at the College of William and Mary, young William Barton Rogers packed his belongings and left for Maryland. Accompanied by his brother, Henry Darwin Rogers, William headed for the small town of Windsor located near Baltimore. William had only a vague idea of what lay ahead, but between the time he arrived in 1825 and his return to Virginia three years later, his life path had become much more clearly defined.[1]

The first thing Rogers did in Windsor was start a small Latin grammar school (the Victorian-era equivalent of a high school), but his interests quickly turned to higher learning. During his years in Virginia, he assumed there'd be plenty of opportunities for educators in Maryland. He was pretty sure that the classical and scientific education he had received at William and Mary would open doors in and around Baltimore. It turns out he was mistaken. "Teaching," he discovered, was "much less profitable in Maryland than in Virginia." He struggled along with his brother until he found an opening at an institution in Baltimore. Leaving his brother in charge of the Latin grammar school, William became an instructor at the Maryland Institute. Through its popular lectures and public exhibits, the small, recently opened institute specialized in "scientific information connected with the mechanic arts, among the manufacturers, mechanics and artizans of the city and state." There William began to show a strong interest in the organization of science programs and their function in the higher learning landscape. The appointment stimulated his imagination about the potential of such institutions.[2]

The instructorship in Baltimore also directed his attention squarely onto science. Rogers taught in the areas of mathematics, physics, chemistry, and astronomy, without any requirements to teach the classical languages. Even at the school he founded in Windsor, the classics stood at the center of his teaching regimen. But at the institute he felt free to prepare lectures in the sciences and to develop research interests in the fields he taught.[3]

As his time and interests shifted toward higher learning and science instruction, Rogers began to take seriously the role of laboratory instruments. He became convinced that "apparatus" played an important role in the advancement of science. It offered scientists new ways to examine nature and to advance their understanding of natural phenomena. But these tools were expensive, and Rogers saw scarcity of equipment as an obstacle to the advancement of knowledge and instruction. "The want of apparatus," he noticed, "has compelled me entirely to omit several subjects in my department." As he faced large audiences of science enthusiasts at the Maryland Institute, the newly installed, twenty-three-year-old instructor became absorbed with how best to present scientific knowledge with the available instruments. When he hit on the right one for the job, he could hardly contain his enthusiasm.[4] "I am at present engaged with the subject of astronomy," he sunnily reported back to family in Virginia, "and have already delivered four lectures upon it, in which I have been much assisted by an admirable [device] which has been loaned to me. It would be difficult to give you an idea of the beauty of this instrument. It was constructed by an ingenious young mechanic in this place a few years ago, and . . . is still of great value in illustrating many important points in astronomy. . . . The instrument affords a clear explanation of the phenomena."[5]

During this period in Maryland, Rogers discovered the areas of interest that cleared a path toward the work of a lifetime; his interests in higher learning, science, and the laboratory remained with him through fundamental transformations in the nation. The United States, by then, was struggling to shed its Old World traditions of college life centered on the classics, traditions with limited opportunities in the way of science. Scientific knowledge had far outstripped the rudimentary offerings at most American colleges and universities, leaving instruction in this area to nontraditional institutions. Science itself, still young and groping in the New World, awaited a group of scholars who could devote themselves to the pressing questions of the day. Amateurs and aspiring professionals alike engaged questions about the parameters of science and the need to reconfigure its fields of inquiry. The nation had developed ever-greater needs and demands for innovations in instrumentation, moreover, whether for the laboratory or the textile mill. Nevertheless, the relationships between the laboratory and science—particularly with regard to instruction or to the advancement of knowledge—interested but few in the academic world. Although Rogers considered his future uncertain while in Maryland, he had found three main pursuits. In these early years the teaching experience turned into a passion for inquiry and the profession as well as an insatiable appetite for ideas about institutes of technology.[6]

First and foremost among the influences that shaped Rogers's life pursuits was his father, Patrick Kerr Rogers. Born in Ireland in the year the American Revolution began, Patrick was first schooled under his own father, then in a local school-house, and later by private tutors. Shortly after turning twenty, Patrick participated in a rebellion against British rule and published antigovernment articles in a Dublin newspaper. Known for his controversial political activities, he fled Ireland for fear of persecution and sailed to America. In 1798 he arrived in Philadelphia and soon after followed an interest in science to a tutorship at the University of Pennsylvania.[7]

Few cities in the New World could've provided as many opportunities as Philadelphia for aspiring scientists of Patrick's generation. The city of fifty thousand citizens still housed the nation's capital. Prominent political and intellectual leaders visited Philadelphia and continued to do so after America's governing center moved to the District of Columbia. One mainstay of the old site was the American Philosophical Society. By the time Patrick had arrived from Ireland, the society had a meeting hall, a collection of natural specimens, a host of apparatuses such as telescopes and other equipment, and a depository for maps, drawings, and models. From its equipment to its leadership the organization reflected a national fervor for the practical over the theoretical. "Knowledge is of little use," noted the first publication from the society, "when confined to mere speculation. But when speculative truths are reduced to practice . . . knowledge then becomes really useful." Most important to Patrick, however, was the newly organized University of Pennsylvania. In the late eighteenth century the state legislature brought the institution into being through a merger that combined the College of Philadelphia and its medical school with the University of Pennsylvania. The most notable beneficiary of the merger was the medical school, which soared in prestige and attracted to its campus students from Europe and a renowned medical faculty.[8]

Patrick worked with colorful, innovative professors at the University of Pennsylvania who left their mark on his scientific and educational thought. He attended the lectures of James Woodhouse, for example, a chemist and physician who founded one of the first professional organizations for chemistry, the Chemical Society of Philadelphia. For seventeen years Woodhouse held the presidency of the association. During that time he also became well known for original laboratory experiments in the areas of medicine, commerce, and industry. Despite his sometimes outlandish claims (that "by chemical agency alone, he could produce a human being," for example), these experiments promoted interest in plant chemistry, chemical analysis, and the use of the laboratory methods for chemical instruction. One observer recalled that "his laboratory was in sundry places perpetually glowing with blazing charcoal, and red-hot furnaces, crucibles, and gun-barrels, and often bathed in every portion of it with the steam of boiling water." His use of the laboratory attracted the attention of many American scientists; Patrick, and his future family, took the idea perhaps more intently than most.[9]

Botanist and geologist Benjamin Smith Barton also had an impact on Patrick's career at Pennsylvania. "If, in the course of my life," he wrote to his professor, "I may enjoy any happiness from my attachment to the sciences . . . it must be acknowledged . . . the result of your example, instruction and benignity." Barton's most recognized work, *Collections for an Essay Toward a Materia Medica of the United States* (1798), appeared the same year that Patrick arrived in Philadelphia from Ireland. The study compiled a list and description of American medicinal plants based on original observations and inquiry. While he studied under Barton, the professor also published *Fragments of the Natural History of Pennsylvania* (1799), a work that inspired many of his students to pursue the newly forming field of geological studies. Patrick numbered among those who took an interest in the field (an interest he would pass down to his sons). Pennsylvania had long recognized Barton's expertise in the discipline and impact on his students. To ensure that he'd stay at the institution, officials established the first chair of natural history in America.[10]

More influential on Patrick than either Woodhouse or Barton was the eminent Benjamin Rush. In the mid-eighteenth century Rush held one of the first chairs of chemistry in America, at the College of Philadelphia. Although he achieved distinction in both medicine and chemistry, his interests extended beyond the two fields. Rush established a reputation as a social reformer by publishing articles on the slave system and organizing the Pennsylvania Society for Promoting the Abolition of Slavery. In addition, his thoughts on education reform figured most prominently in the lives of the Rogers family. In a post—Revolutionary War article in Philadelphia's *Federal Gazette,* Rush outlined his plan for a federal university. Unique to the proposal, he envisioned curricular and pedagogical changes to the dominant classical course of study. "Let those branches of literature only be taught," he argued, "which are calculated to prepare our youth for civil and public life. These branches should be taught by means of lectures, and . . . arts and sciences should be the subject of them." Responding to traditional collegiate models that emphasized Latin and Greek taught by way of the recitation, Rush advocated the establishment of an institution that would place practical and scientific studies taught with alternative methods of instruction at the center of the curriculum.[11]

While studying at the University of Pennsylvania, Patrick met Hannah Blythe, also from Ireland. Hannah came from a free-spirited family that was involved in antigovernment activities in Ireland, providing a strong commonality between the two immigrants. Her father had once held secret ownership of the *Londonderry Journal,* known for its controversial articles. But by the late eighteenth century both her parents had died, and, along with her sisters, Hannah left for the United States. Arriving in Philadelphia, the Blythe sisters were received by a cousin who had fled years earlier. Soon after Hannah settled in America, she became engaged to Patrick. In keeping with their heritage, they married in Philadelphia's Presbyterian church in 1801. The following year the couple had their first child, James Blythe Rogers.[12]

Having started a family and graduated from the university in May 1802, Patrick formed a medical practice in Philadelphia. Although he'd found his professors inspiring and was drawn toward academic life, he saw medicine as the only viable career. Around the time he received his degree, few undergraduate programs kept faculty positions in science as stable as Harvard's Hollis chair or Princeton's chemistry professorship. In fact, most colleges showed only mild interest in nonclassical studies. Practicing physicians, meanwhile, enjoyed relative stability and a newly acquired sense of professionalization. Although only three American institutions—Harvard, the University of Pennsylvania, and King's College (later Columbia)—had medical schools at the turn of the nineteenth century, twenty-six new schools opened over the following four decades. This expansion in medical education contributed to a distinction between those in the mainstream and those without formal education in the field. The medical degree offered Patrick entry into the world of mainstream medical practice, which at the time still included such primitive treatments as bloodletting and purgative methods for even mild illnesses. Entry, however, didn't necessarily secure a livelihood. Regardless of educational background or treatments employed, Patrick, like most of his peers, depended on establishing a reputation among clients for a successful career. He had firsthand experience with the challenges of the occupation, especially after a brief trip to Ireland. "In the year of 1803," he reflected, "I was engaged in full business in Philadelphia as a physician, and the products of my practice were more than equal to my current expenses." The same year, unfortunately, his debts ballooned after he tended to his father's funeral in Ireland. When he returned to his practice in Philadelphia "to make a second beginning in the same place," he recalled, "I was never able to procure a share of business equal to the expenses of my family, however moderated." Whatever reputation he had cultivated in the community perished during his brief absence.[13]

The practice dissolved soon after Patrick returned from Europe, and he looked to other opportunities to recuperate his losses. One unusual venture consisted of an attempt to start a medical library for the city of Philadelphia. After investing a substantial portion of the family's savings, the library became a short-lived experiment and ultimately failed. During this trying period in the family's history, William Barton Rogers was born, on December 7, 1804, followed by a third son, Henry Darwin Rogers, four years later. At that time the Rogers family sank deeper into debt, and

Patrick's creditors recommended that he do what many other Americans had begun to do: pull up stakes and move to another city in search of better opportunities. The credit lenders suggested New York and Maryland, and the troubled physician settled on the latter to give medicine another try.

By 1812 the Rogers family had moved to Baltimore and opened an apothecary shop. Their prospects increased moderately as Patrick added a series of successful public lectures to his schedule. To a large extent it was his lecturing abilities that won him membership in the Hibernian and Maryland Medico-Chirurgical societies. While his professional opportunities increased, so, too, did the size of his family with the birth of Robert Rogers, the fourth and final son, a year after their move to Baltimore. Six years later Patrick, still seeking greater economic stability, applied for a professorship at the newly organizing University of Virginia. In a letter of application for the appointment he commented that during his years in Philadelphia he had "delivered several courses of lectures on chemistry and natural philosophy in Philadelphia, some of which were attended . . . by the director of the mint, Robert Patterson, and several of the professors of the University of Pennsylvania." Patrick hoped the opening at the University of Virginia would allow him to pursue teaching and research in natural philosophy and mathematics. Certainly, the educational ideals of the new university, resembling the science-promoting notions of his former mentor, Benjamin Rush, must have interested him. Yet his strong appeals to the university yielded nothing except a letter of rejection from the founder, Thomas Jefferson. In the same year, however, Patrick received an offer to join the faculty at the College of William and Mary in Williamsburg, Virginia. With little hesitation the Rogers family collected their possessions and headed south.[14]

Not long after they arrived in Williamsburg in 1819, the family was already considering returning to Baltimore. Patrick valued the professorship of chemistry and natural philosophy at the College of William and Mary, yet, as he wrote to his colleagues in Maryland, life in the small college town left much to be desired. Patrick came into conflict with the institution's policies and had difficulty with what he perceived as a stale intellectual climate. The two eldest of Patrick's four sons, James and William, enrolled and also found the transition difficult. James decided to leave after two years, returning to Baltimore to complete a medical education. William, however, stayed on for six trying years before following a similar path. The immediate difficulties the Rogers family faced came from William and Mary itself, but the problems reflected a deeper, more pervasive obstacle to higher learning in the rural South: isolation.[15]

The contrasts between urban and rural life in early-nineteenth-century Baltimore and Williamsburg provide an explanation for the challenges the Rogerses experienced. By 1819 Baltimore, with its sixty thousand residents, had developed into a center of international commerce. The European wheat trade favored the centrally located city and its ports, as it had since the mid-eighteenth century. With the rise in merchant trading came a major shipbuilding industry, an artisan community, and a proliferation of shops, banks, and insurance companies. As an emerging center of commerce, Baltimore began to compete with nearby Philadelphia. Not to be outdone in either education or science, the state of Maryland granted the request of three Baltimore physicians to establish the College of Medicine in Maryland in 1807, an institution that went on to "promote medical knowledge." Five years after its founding, the college added a Faculties of Divinity, Law, Arts and Science and changed its name to the University of Maryland. The emergence of the new institution mirrored the expansion occurring in Maryland's largest city, growth that had attracted nearly a third of the state's population.[16]

Williamsburg, on the other hand, had a distinguished past but an uncertain future. The city of two thousand inhabitants had been home to Virginia's Assembly, giving rise to the legislative careers of George Washington and James Madison. Williamsburg's oldest establishment, the College of William and Mary, derived its name from the monarchs of England who granted the charter in 1693. After Harvard College, William and Mary became the second institution of higher education founded in the colonies. The Virginia college originally proposed to open "a certain Place of universal Study, a perpetual College of Divinity, Philosophy, Languages, and other good Arts and Sciences." It began by providing a seminary for the Church of England in the colonies and offering a general education to the social elite of Virginia. Among its alumni William and Mary could claim Thomas Jefferson, James Monroe, and John Tyler. Before challenging the college with a rival institution, Jefferson declared, "I know of no other place in the world, while the present professors remain, where I would so soon place a son." The professors ultimately left, however, and problems facing the once lively political center of Virginia increased when the state's capital moved to Richmond. Without a significant base for commercial or political life, the Rogers family believed the small college town had simply fallen off the map.[17]

The Old South had no special monopoly on the challenges of rural life. Certainly, communities in other regions, especially in the Midwest and further westward, were equally removed from the activity in the Northeast. Yet intellectuals across the South were particularly bitter about their sense of exile. Literary figures such as antebellum novelist William Gilmore Simms grew despondent over what he viewed as a drab intellectual life in the region. "I have never known what was cordial sympathy in any of my pursuits among men," bemoaned Simms in South Carolina. He longed for the fellowship and support of others with similar concerns. In a self-described "Sacred Circle" Simms finally found a small community to console him. As a group, numbering five social critics in all, they deplored the lack of cultured pursuits in the South. One member of the circle, politician James Henry Hammond, even while governor of South Carolina, considered himself intellectually "as solitary . . . as if I were in the great Sahara." Edmund Ruffin, another member, spent most of his career advocating scientific reforms in southern agriculture but had little impact on the region. He believed that "the great evils which serve to prevent agriculture from being prosperous in Virginia may be summed up in a single word, ignorance." Political economist George Frederick Holmes, who floated in and out of academic positions across the South, looked for consolation in the circle as well. None of the institutions—including William and Mary, the University of Mississippi, and the University of Virginia—had provided him with a satisfying community of scholars, leaving him a veritable "alien on a desert shore." Lawyer and novelist Nathaniel Beverly Tucker, the final member, also taught at William and Mary with little intellectual companionship. He viewed the condition in the region with great scorn, especially when he commented that even Robinson Crusoe had been "hardly more completely isolated than I."[18]

At William and Mary isolation made it difficult for the institution to attract and retain scholars like Patrick Rogers. In natural philosophy the two professors before him both left Virginia for more stimulating climes. Thomas P. Jones held the professorship until 1817, before moving on to a career at the Franklin Institute in Pennsylvania. That year the contentious president of the college charged that Jones's lectures lacked a "scientific" quality. The attacks from the president, coupled with the area's want of intellectual activity, Jones complained, kept Williamsburg in a "humiliating condition." Well-known chemist Robert Hare replaced Jones, but for only a year. By 1818 Hare accepted a professorship at the University of Pennsylvania and went on to establish a prominent career in chemistry at the Medical School. For the following session William and Mary elected Patrick to fill the vacant chair. He missed the kind of activities that he left behind in Baltimore—the circuit of public lectures, the Medico-Chirurgical Society meetings. Ruminating over the character of life at the college, one of his colleagues described the area as a "sad place of solitude and exile."[19]

William Barton Rogers first experienced such intellectual isolation as a student at William and Mary. As with his father and the Sacred Circle, he thought of himself as somewhat disconnected from the life of his peers. Unlike the less-prepared students he would come to know, William had studied for college under his father, who provided a home education for his sons. Patrick had good reason for doing so. The geographic distribution of people in the South, widely scattered across large regions, made it nearly impossible to organize statewide school systems. Local public systems rarely emerged outside of major southern cities, such as Nashville and New Orleans before the Civil War. Instead, private, sometimes fly-by-night institutions appeared here and there to meet specific needs as they cropped up. William later recalled the details of his early schooling, noting that "with the exception of a short period . . . we never spent any of our afternoons in schools." Happy with this alternative to the "drudgery" his playmates experienced, William recognized that he and his brothers had gained a "thoroughness of our knowledge on all the subjects which we studied."[20]

Few of his colleagues, however, had been given similar opportunities, and for the most part were unprepared for academic life. Given the limited options available for preparatory schooling at the time, colleges and universities received students ready for high school rather than higher education. Because tightening requirements would prove fatal to most small institutions, undergraduates across the southern states received a course of study that matched their level of preparedness. Rogers's home education colored his observations of Virginian college life, especially the life of his peers. Toward the end of his first term at William and Mary, he mentioned to one of his brothers that "with the exception of about eight, there was perhaps never an assemblage of young men so totally destitute of genius and so miserably deficient in understanding. Yesterday . . . Dr. Smith inquired of a student what was the nature of material substance, the answer was 'one which affects our senses and exerts reason!' Father asked the same person for a definition of a solid; after much hesitation, a good deal of muttering, and abundance of broken sentences, the gentleman answered with great philosophical gravity that it was 'A-a-a body which was a solid.'"[21] Rogers was ambitious and competitive and, at least in natural philosophy, ahead of his peers. But his competitive nature spilled over into other subjects as well. He developed a strong secondary interest in styles of rhetoric and oratory through recitations in the classical languages. By the age of seventeen the community selected him to give a "Virginiad" oration commemorating the founding of Jamestown, Virginia. The Rogers family took pleasure in the attention they received when state newspapers published young William's speech.[22]

While the lack of a scholarly spirit on campus bore down on him, Rogers also grew weary of cultural isolation. For the period he might have been asking much of the little town. In the antebellum South only two centers of commerce, Charleston and New Orleans, held promise for cultural activity. Coastal access to the mercantile economy had not only provided the cities with opportunities for the importation and exportation of goods but also of ideas. Charleston could boast a Library Society and a Literary and Philosophical Society. The former became the famed Charleston Museum, one of the oldest repositories of natural history for the region. The latter promoted an interest in "every department of the arts and sciences." New Orleans developed similar cultural organizations during the period, with its Lyceum of Natural History and Society of Natural History and Sciences. Although literary societies and museums began to appear in the two cities, the rest of the South experienced cultural isolation. As the author William Gilmore Simms declared, "The South don't care a d——m for literature or art." Williamsburg, apparently, was no exception.[23]

Rogers lamented the lack of academic cultural activity and observed the "foolish" character of life on the campus. Taking himself and his work perhaps too seriously, he described the behavior of the students at William and Mary as carnivalesque in their obsession with "feasting, dancing, and music." "Students," he stated, "are more occupied in anticipation of the pleasure that one evening will afford them than in preparing themselves for the appropriate chair of examination." While he linked cultural isolation to problems with the extracurriculum, he associated the same with gaps in the curriculum. As a student, Rogers faced several challenges while assisting his father collect materials for lectures and science demonstrations. Few of the materials and equipment they needed could be purchased locally. Rogers wrote to his brother James in Baltimore: "I wish you could learn whether Doctor [Elisha] De Butts has yet prepared Iodine or potassium . . . father is unable to present to class for want of [it] . . . and the apparently necessary for preparing the latter cannot be obtained anywhere in Virginia." Sending away for virtually every chemical and every apparatus, their frustrations mounted.[24]

Rogers's father looked for ways to enhance the experiences of his sons and counter the limited intellectual and cultural resources of Williamsburg. For William this meant being allowed to participate in research projects. Upon completing a mathematics text, for example, Patrick wrote to Thomas Jefferson that some of the problems and their explanation "are by my second son who is now in his 20th year and has a very extraordinary passion for physico-mathematical sciences." In the same letter to Jefferson he mentioned, "I intended to have it [the text] sent to you last year but was induced to defer doing so from the expectation that, I should before now, have found convenient to get the diagrams engraved: the state of this institution [William and Mary], however, does not encourage me to incur this expense." Indeed, he feared that the college would hardly be able to compete with Jefferson's proposed University of Virginia. "There is something in the organization of William and Mary," Patrick told the former president, "which independently of its location or other permanent disadvantages, must forever prevent it from being prosperous or successful. . . . I am inclined to think that when [the university] goes into operation we shall scarcely have occasion to open the doors of the old College. Even at present there is no reputation to be acquired here, and no encouragement to activity or zeal."[25]

Any bright expectations the Rogerses might have had for the College of William and Mary continued to dim. External signs of the institution's decline began to appear and certainly didn't help morale. Visitors passing through Williamsburg would comment on the sad condition of the campus. One traveler described the scene as "the ruins of William and Mary College." "It has been very much neglected," continued the description, "and will soon go quite to ruin. The steps are mostly out of the place. Some of the windows are entirely broken out and most or all of them more or less broken, some not having more than three panes of glass in them. The cellar is used for a barn, and the building has more the appearance of a gaol in ruins than the remains of a college." Students wanted to leave the discouraging scene as much as the faculty did. Enrollment fell from an average of thirty-four to a total of six students in 1824. That year President John Augustine Smith, supported by two faculty members, proposed to move the last vestiges of the institution to Richmond in a final effort to survive. The idea of reviving the college in a more prosperous economic and social environment was repeated often by advocates for the move. They hoped that enrollments would rise in a more densely populated city. After gaining approval from the Board of Visitors, Smith sought funds from the state legislature to finance the effort. A steady stream of appeals from William and Mary, however, was met with a yawn. Legislators had little interest in funding the transfer, leaving the institution in Williamsburg as isolated as ever.[26]

To Rogers the college's future looked bleak, and the following year he left for Baltimore with his brother Henry. The two joined their oldest brother, James, leaving the youngest behind in Williamsburg. Despite William and Mary's problems, their studies under their father had prepared them for scientific lives; individually and through collaborations, they embarked on careers in science that gave rise to the fraternal circle known as the four "Brothers Rogers." They greatly influenced each other's social, scientific, and educational views, and their personal and professional life histories help shed light on the development of William's character and persona.[27]

James Blythe Rogers had the most turbulent entry into the scientific community. After a brief stint at William and Mary, he began medical studies at the University of Maryland, where he received an M.D. degree in 1822. His first attempt at establishing a medical practice with a colleague was a dismal failure and prompted him to accept a position as a superintendent of a large chemical factory. Unsatisfied with factory life, he sought other means of employment, including tutoring and intermittent lectureships at the Maryland Institute. Incessant pleas to his father for support reveal the tenuousness of his teaching arrangements. At one point James entertained the idea of providing medical assistance to a colony of freed slaves at "Cape Mesurado" on the west coast of Africa. Although the opportunity never materialized, he ultimately journeyed to another frontier, Ohio. For almost five years he taught in the medical department at Cincinnati College as professor of chemistry, followed by positions at the Philadelphia Medical College, the Franklin Institute, and finally the University of Pennsylvania until his death in 1852. His professional and scholarly accomplishments included assisting in the organization of the American Medical Association, memberships in the American Philosophical Society and the Academy of Natural Science of Philadelphia, and publications in chemistry.[28]

Henry Darwin Rogers, the third son, also attended William and Mary for his undergraduate studies until 1825. He spent the next several years in Maryland employed in various commercial and educational occupations until he was offered a teaching position at Dickinson College in Carlisle, Pennsylvania. After only a year, however, the trustees dismissed Henry for his reform-minded views on science and its role in traditional education. In a controversial article he argued against the dominance of the classical curriculum and advocated alternative methods of instruction. The arguments had much in common with the views of his father's professors at Pennsylvania. After leaving Dickinson, his interest in reform continued unabated. "The true struggle for human liberty," Henry remarked, "is in the field of education, by the pen and through the press, it is in the hall of knowledge and on the leaf of science." Although he was active in promoting such beliefs, his most significant contributions were in geology. His dismissal from Dickinson allowed him time to conduct scientific research in Europe. While there, Henry formed lasting personal and professional relationships with some leading British geologists. His reputation as a natural historian swelled when he returned to the United States after a few years, and he was commissioned to lead the state surveys of New Jersey and Pennsylvania. The survey work and other research endeavors yielded numerous publications and launched his career as one of the first professional geologists in America. He subsequently returned to Europe and accepted the Regis Professorship at the University of Glasgow in 1857, a position he held until his death nine years later.[29]

The youngest brother, Robert Empie Rogers, also enrolled at the College of William and Mary. There he started a lifelong interest in chemistry, which continued at the University of Pennsylvania. Robert graduated with an M.D. degree in 1837 but maintained no serious interest in becoming a practitioner. Instead, he accepted a professorship at the University of Virginia, where he kept up his chemical research. When his brother James died, leaving open the chemistry position at the University of Pennsylvania, Robert was called to fill the vacancy. He accepted and later spent twenty-five years as dean of the faculty. After his tenure at Pennsylvania, he left for a professorship in chemistry and toxicology at the Jefferson Medical College in Philadelphia, where he stayed until 1884, when he became ill and died. Although less prominent than his brothers, Robert had an impact on the development of scientific education, conducting numerous chemical experiments and advocating the use of the laboratory for instruction in higher education.[30]

William Barton Rogers, the second son, interacted frequently with his brothers in an ongoing dialogue that contributed to the shaping of his educational and scientific thought. His wife, Emma Savage, would later recall that "the lives of the three brothers . . . occupied so large a share of his thought and affection." Much of the correspondence between the family members encouraged one another to stay current with advances in science and collegiate reforms. Early in William's academic career, for example, James made sure his younger brothers made the most of their scientific studies. He encouraged William to develop a critical approach to science that went against the rote memorization methods of the traditional recitation: "I now sit down to write you a short letter, in which you may not calculate on anything new, except a new and in my opinion a rather singular opinion advanced by Dr. [Elisha] De Butts, which he delivered this evening, one which I think is wholly unsupported by any evidence." When William ended his studies, he took this critical approach and his research interests to Baltimore.[31]

As mentioned earlier, William, with Henry's assistance, envisioned opening a Latin grammar school in the town of Windsor. Their efforts organizing, designing, and managing an educational institution paid off in the fall of 1826, when the school opened. They had modest hopes; to them the school was a brief detour along a path toward a career, one that would "in a few years" allow them "to obtain a profession and begin the practice of it." They soon realized that it wouldn't work out that way. The modest income could not "expect to make much more than a support in our present condition." "The profits of the school," Rogers complained to his father, "would be sufficient to satisfy one of us, as it would enable him to lay up something for the future." William and Henry couldn't both stay at the small institution and also save for further studies. William, therefore, began looking for work in Baltimore and found a lectureship at the Maryland Institute that suited him.[32]

The Maryland Institute, founded in 1825, was modeled after Philadelphia's Franklin Institute. Both of them sponsored public lectures and demonstrations in the sciences. Rogers started his lectureship at Maryland two years after its founding and there learned of profound changes occurring in American society and the changing climate of educational discourse concerning science. In his introductory address at the institute Rogers reveled in "the usefulness of popular courses of scientific instruction" and noted that similar courses had become more common in America by this time. "Of late years," he stated, "the public mind, both in this country and abroad, has been much interested in the subject. In many places institutions calculated to render useful science attainable by the mass of society have been established; and such is the growing impression of their value that their number continues yearly to increase." Increasing attendance at the institute's lectures gave him further reason to believe that there was indeed a growing interest in American science. Although he taught with few instruments and even fewer opportunities for laboratory sessions, he relied mostly on his voice, the blackboard, and a few demonstrations to attract an audience. His enthusiasm for science and practical innovations, making use of the available "apparatus" and developing his public speaking abilities, opened doors to promotion for him at the institute.[33]

His reappointment from temporary lecturer to professor in Baltimore ultimately depended on a single condition. If the Maryland Institute would purchase the resigning professor's equipment, Rogers would assume the chair. Dealings between the institution and the departing professor broke down, however, and Rogers sulked to his father: "Had they purchased it, I would certainly have been appointed. As it is, I presume no appointment will be made." In the end he never received the promotion at the Maryland Institute, but he gained valuable experience in the conduct and organization of an institute of science and practical studies.[34]

The instructorship lasted only one session, but the institute trustees provided William Rogers with another opportunity. They asked him to develop a plan for a feeder school that would be directly affiliated with the institute. Colleges and universities of the antebellum period often established classical schools and academies to prepare students for higher learning. But the Maryland Institute wanted something different: a school with a greater utilitarian aim than most preparatory schools of the period. In April 1828 Henry informed his father that "William is at present engaged in maturing a scheme for the regulation of the school, to be offered to a committee of managers for their approval." The following day Rogers submitted his fully formed plan for the new school.[35]

His plan had five principal elements. First, he outlined the comparatively low cost of tuition and the items covered and not covered by the price of instruction. Second, he established a set of minimum entrance requirements. All new students needed to show proficiency in spelling, reading, writing, and "arithmetic computations at least as far as the rule of three." The total number of students for the school, as determined by a third point in his plan, could not exceed fifty. Rogers did not mention the number of staff members in his statement, but he had at least himself and his brother Henry in mind. Fourth, the plan ruled out any offerings in the ancient languages. "Classical studies," he made clear, "are not within the scope of the school." And the fifth and most significant element described the mission of the school. The new institution's purpose was to provide students formal preparation for "mechanical and mercantile employments." To accomplish this goal, the curriculum focused on mathematics, geography, surveying, navigation, and English composition. Within a month after Rogers submitted the proposal, the Maryland Institute prepared to open the school with Rogers in charge of its administration and teaching, providing the twenty-three-year-old with his second experience establishing and leading an educational institution. His early exposure to planning, teaching, and administration greatly advanced his interest in the promotion of practical education.[36]

During the first week of May 1828, shortly before the school opened, William and Henry went to Philadelphia to examine a feeder school affiliated with the Franklin Institute. William believed the trip would give him more ideas to consider. What he discovered at the school probably surprised him. He knew that the Franklin Institute resembled the Maryland Institute in several ways. Both provided popular lectures, a science curriculum, and practical courses in the "useful arts." But the institute in Philadelphia had opened a classical high school two years before their visit. Over the three-year course of study at the school, students "took Greek and Latin every year, three years of mathematics and French, two years of Spanish and drawing, plus courses in history, geography, political economy, astronomy, natural philosophy, chemistry, bookkeeping, and stenography." The curriculum Rogers had in mind for his school differed markedly. He saw little need for the classical languages for his students and believed his institution should have a more specific, practical focus. While the Philadelphia school prepared students for classical colleges, Rogers offered an entry into the world of science and practical knowledge.[37]

Rogers's scheme apparently filled a need. Two dozen Baltimore students signed up from the start. He invited Henry to join him in the teaching duties, and both instructors soon found the load burdensome. "Henry and I have found our engagement very fatiguing," he commented. "We have recently instituted a plan in the school which enables us to relieve each other on alternate days." Despite the labor-intensive routine, Rogers took pleasure in the upsurge of interest in practical education. In Baltimore he witnessed a popular excitement over emerging technologies and science. A grand procession of the state's officials and citizens, for instance, filled the city in celebration of the construction of a railroad. Rogers also found his lec-

ture hall crowded with curious listeners and marveled at the rapid growth of institutions across the nation that provided courses for science enthusiasts.[38]

In August 1828, Rogers's time in Maryland came to an abrupt end, however, when
his father, Patrick, died. Traveling on the way to visit his sons in Baltimore, Patrick
fell to malaria, which had also taken Hannah, his wife, to her death eight years earlier. Following the tragic summer month, William received an offer from the College of William and Mary to fill the chair previously held by his father. By September he had submitted letters of recommendation from colleagues and, shortly
afterward, secured the position. He received more than recommendations from his
colleagues; he also got advice. Some viewed the position as an excellent opportunity.
Others were not as sanguine. The college by this time had an uneven reputation,
and their advice raised such concerns. One advisor warned that perhaps "your ulti-
mate advancement would be more promoted by your remaining here. They state
that there is now opening in this country an extensive field for highly respectable
and lucrative exertion in the growing spirit for works of internal improvement de-
manding the superintendance [*sic*] of scientific men." Rogers, well aware of exciting
possibilities in Maryland with internal improvement projects, stood at a cross-
roads.[39]

CHAPTER TWO

Tenure in the Tumult

WILLIAM BARTON ROGERS had a decision to make: Would he stay at the Maryland Institute or return to his alma mater as a faculty member? Most candidates at the time wouldn't have given it a second thought. Moving from a little-known institute to a well-recognized college made the most sense. But Rogers's experiences as an undergraduate at the College of William and Mary complicated matters for him. Isolation in the region had cast a cloud over his student years; the problems associated with that intellectual and cultural isolation were still fresh in his mind. He also expected difficulty teaching at Williamsburg. Having witnessed the institution's enrollments drop to single digits cooled his interest in the position. What's more, Rogers knew that students who did manage to enroll had little or no background in the sciences. Leaving the instructorship at the Maryland Institute would mean fewer opportunities for preparing the kind of popular and advanced lectures he found rewarding. On a personal level going back to Virginia would mean revisiting family losses. Both of Rogers's parents had died from malaria emanating from the swamps surrounding Williamsburg.[1]

Nevertheless, the young scholar accepted the position, and from the scant records of the period we can only speculate about the reasons why. In his inaugural address, infused with a eulogistic tone that mourned his father's death, Rogers disclosed some of his reservations over returning, calling himself an "inmate of the halls" in which he had once studied. Yet he may have felt honored with the opportunity to fill his father's former role at the college; after all, it was from Patrick Kerr Rogers that William had received extensive instruction in natural philosophy, natural history, and mathematics. Moreover, while William and Mary teetered on the brink of collapse during the previous decade, it could still claim a distinguished legacy in American higher education. With Rogers seeking to enter the academic world, the science professorship there must have had a strong appeal. That distinguished chair came with a stipend that was greater than the one he had collected from the Maryland Institute. For years his father had provided a stable source of support to the family, and now William would quickly offer the same for his brothers. A sense of duty and responsibility doubtless figured into his reasoning.[2]

When Rogers made his final decision to return to William and Mary, he began what turned into a twenty-five-year career as a southern professor. His first seven years in Williamsburg, followed by the remaining eighteen at the University of Virginia, coincided with a tumultuous era for the region and the nation. It was through these academic and leadership roles in Virginia that he came to experience a microcosm of this tumult.

Rogers identified two basic challenges to campus life in the South, both of which he closely associated with the problem of slavery and both of which persisted throughout his southern career. The first challenge had to do with the character of social violence in the region. While northern and southern communities faced roughly the same number of cases of violent behavior, the incidents in each section differed largely in kind. For the most part northern mobs tended to damage property, while southern mobbings more frequently ended in personal injury. Likewise, campuses in both sections may have shared similar institutional features, but they differed substantially in cultural terms. The organization of higher education, with its curriculum, methods of instruction, and the daily routines of students and faculty, proved to be virtually identical across regions. Even innocuous student pranks, such as ringing the chapel bell in the early hours of the morning or burning an infamous professor in effigy, appeared at nearly all institutions of the nineteenth century. But as for frequency of life-threatening incidents by student mobs and rioters, southern colleges experienced a disproportionate share.[3]

That Rogers associated the culture of violence with slavery was nothing new. Virginia's Thomas Jefferson had long before argued that the South, when compared with northern states, maintained a distinctively violent culture. Jefferson, himself a slaveholder, blamed the hostile environment on the "peculiar institution." "The whole commerce between master and slave," he wrote in the late eighteenth century, "is a perpetual exercise of the most boisterous passions, the most unremitting despotism on the one part, and degrading submissions on the other. Our children see this, and learn to imitate it." Assuming the master's role, the child "puts on the same airs . . . [and] gives a loose to his worst passions." Rogers made similar observations in relation to campus life. In particular, he found that southern students often lost control of their passions, especially when southern professors failed to view students as sons of a master class. Slavery had made students sensitive to any orders or commands or demands made on them by college faculty. If students interpreted a demand as a breech in the slave society's code of honor, faculty could expect a fiery reprisal.[4]

Rogers's earliest encounters with such behavior occurred at the College of William and Mary. Certainly, William would have recalled that, during his student years, he'd seen his father's life threatened over a trivial comment. The incident began when Patrick Kerr Rogers reprimanded a student named John A. Dabney for whispering in class. Embarrassing Dabney before his peers, Patrick remarked that "such conduct was utterly inconsistent with the character of any Gentleman in polite Company." Sensitive to being ordered about and having his honor put into question, Dabney resorted to violence to settle the issue. After class he sought out the professor and threatened him with a menacing stick. The student told Patrick that "his gray hairs only, protected him from the Punishment which his Conduct merited," at which point the two began to scuffle. Despite the life-threatening gestures, Dabney only received a temporary suspension from William and Mary. To Patrick's dismay the student's uncle turned out to be the president of the college.[5]

After William replaced his father at the college, he became subject to similar student passions. Indeed, Rogers found himself confronted one evening in 1832 by Charles Byrd, who brandished a stick in one hand and a loaded pistol in the other. The enraged student, recently disciplined for riding a horse inside a campus building, addressed Rogers in "the rudest and most insulting language," while failing to declare the exact cause of his excitement. When the professor asked the reason for the "epithets" of "abusive and threatening language," the student replied that he "demanded satisfaction" from Rogers. In much the same way that a master abused a slave, Byrd "cried out . . . that he had a mind to cowhide" the professor. Rogers managed to return to the campus house where his apartment was located, but Byrd followed. Finding his target in the house, Byrd held the pistol within a foot of Rogers's heart. Again Byrd demanded satisfaction, this time with the pistol cocked and a finger on the trigger. Rogers described the student's rage as "almost amounting to insanity," in which every moment he "expected to be shot." Eluding the attacker for a second time, the professor fled to his apartment with a friend and locked the door. In a passageway outside Byrd slammed against the door, demanded entrance and satisfaction from Rogers, and swore to shoot the professor and his friend. The incident ended when Byrd decided to leave the building. Following the attack, the faculty voted to forward the case to "the Prosecutor for the Commonwealth" for legal action against the student.[6]

While dealing with campus violence, Rogers came up against a second challenge, which he called "illiberalism," or what historians have depicted as intolerance and anti-intellectualism. Most studies on the period have defined this rise of intolerance in terms of the demise of liberal philosophy, as in the notion of "natural rights," the ideals that provided for freedom of expression, freedom of religion, and freedom of ideas. Thomas Cooper captured the sentiment at South Carolina College when he told his students that "rights are what society acknowledges and sanctions, and they are nothing else." The shift came about as the South began to respond to intense pressure from the North over the slavery issue in the decades before the Civil War, a shift fueled by the Missouri Compromise of 1820, the tariff crisis of the 1830s, the Nat Turner Rebellion, and the appearance of William Lloyd Garrison's newspaper, the *Liberator*. By revising the notion of rights, southerners could justify suppressing debate and diversity of opinion regarding plantation practices.[7]

Rogers's career in Virginia coincided with this changing political climate, providing him firsthand exposure to what southerners believed, how they felt, and what they argued. One of the most galvanizing ideas of the antebellum South spread from his own campus. While on the faculty at the College of William and Mary, his colleague Thomas R. Dew published the first systematic defense of slavery in a *Review of the Debates of the Virginia Legislature*. Others, to be sure, contributed to the cause; Josiah Nott, for example, developed scientific rationales for slavery, and James H. Hammond provided political authority behind the arguments. But Dew spurred the interests of the southern intellectual class. His treatise on proslavery thought maintained a lasting influence until the start of the Civil War. To a large extent the work developed from the issues raised in the Virginia Convention of 1829–30.[8]

The convention, which marked a turning point in Rogers's concerns about the South, met in Richmond to consider amendments to the state constitution, which remained unchanged since its adoption in 1776. At the center of the debates stood the relationship between slavery and the apportionment of political power in the state. Some factions argued for a distribution of power based on population only, meaning that slaves, or "property," would not be counted. Because this proposal would undermine the power held by elite plantation owners, other factions responded with elaborate defenses for the use of slavery in the apportionment formula. In the end the slave interests succeeded in thwarting calls for representation based on population only. The convention revealed the dominance of the slaveholders' political strength as well as their ability to squelch reform.[9]

Rogers also attended the convention, but he came away from the Richmond meeting questioning the proslavery tenor of the debates. In the last days of the session he reflected on the relationship between the defense of slavery and intellectual life in the region. For Rogers, unlike his colleague Thomas Dew, the success of the slaveholding interests held foreboding signs. Rogers went to the Virginia Convention to witness "the proceedings of 'one of the most August assemblys [*sic*] which has ever convened in our country.'" Yet after arriving at the convention to see "all the iminent [*sic*] talents of Virginia . . . constellated together," he declared: "Oh fame how often is thy trumpet stolen by *party* and blown by *prejudice* and *folly*! . . . I have been greatly disappointed in the 'assembled wisdom of Virginia.'"[10] Rogers became convinced that slavery and its defense had a detrimental effect on political and academic life. The intellectual energies of the state, he decried, "have been *misapplied*. They have not been directed to the investigation of the best modes of elevating the *moral nature* of our citizens, of dispensing *truth* in all its purifying and enobling [sic] influences through every section of our state and of establishing that foundation of knowledge upon which every *permanently-good* superstructure in government must be raised." Rather, he continued, "they have been devoted with all the energy of selfish passions, to the most futile energies to balancing and counterbalancing local interests, and local prejudices." The South's defense of slavery, he believed, would contribute to further decline, violence, and intolerance.[11]

Rogers's statements about the convention came as close as he dared, during the antebellum period, to criticizing slavery. For as frustrated as he might have been about its impact on intellectual life, he remained silent about slavery's legitimacy and made no public pronouncements that might antagonize the system. If that silence obscures the nature of his beliefs, the lives of immediate family members provide some insight into his political mind. His father, Patrick, had studied medicine under abolitionist Benjamin Rush at the University of Pennsylvania; Rush had a considerable influence on Patrick's scientific education, and it's likely that he passed along some of his social concerns as well. The careers of the Rogers brothers suggest that such concerns also carried on from father to sons. William's eldest brother, James Blythe Rogers, considered practicing medicine in an African colony of freed slaves at the start of his career. Henry Darwin Rogers, William's younger brother, followed for a time the teachings of Frances Wright and the Owens family, particularly their reformist views, which included abolitionism. The youngest brother, Robert Empie Rogers, often complained of social and cultural obstacles to intellectual life in the South. When asked to return to the region, Robert wrote to William: "Since you as well as I think [I] shall scarcely have the same opportunities of improvement there I am doubtful as to the expediency of leaving this [New York] for Wmsburg. . . . The society too in which I should be thrown is the kind I little relish." None of Rogers's brothers had much in common with the South's politics, and they eventually moved out of Virginia. They went on to accept, or be considered for, science professorships at the University of Pennsylvania (James and Robert) and Harvard University (Henry). Of all the members of the Rogers family, William stayed in Virginia the longest. All the while he privately expressed sympathies for northern social and political developments while never openly challenging southern ones.[12]

Guided largely by practical concerns, he came to lead a double life, with one foot in the North and the other in the South. Francis Lieber, political economist at South Carolina College at the time, followed a similar path. During his tenure in Columbia, Lieber remained vague about his public views on slavery for fear of losing his position or, worse, inciting a mob. He feared even that his visits to New England provoked suspicion, picturing his critics wondering why "he always goes in vacations to the North." As a result of the sectional tensions, he felt he had to silently "cogitate a philosophy of freedom in the land of slavery." While Lieber complained that slavery left the region without "a breath of scientific air, nor a spark of intellectual electricity," he managed to appease his colleagues during his time in South Carolina. Rogers, likewise, held northern sentiments while avoiding prickly topics with his Virginia colleagues. Arch defender of slavery Thomas Dew wrote unwittingly to Rogers, who at the time had left William and Mary, that "it makes me sad indeed to write to a friend with whom I have spent so many happy hours, and laboured so many years in our old college. I miss you exceedingly." Friends in the North understood a different side of Rogers. When a position became available in Philadelphia during the mid-1830s, they knew he would be interested. "I owe many kind thanks to Dallas," commented Rogers about Alexander Dallas Bache, later superintendent of the United States Coast Survey, "for this evidence of his friendly regard, and I hope he will feel assured of the grateful pleasure with which his proposal affected me." The position offered Rogers a way to join his colleagues in the Northeast.[13]

Still, Rogers deliberated. He knew that developments in the Virginia legislature, particularly an interest in the state's precious mineral deposits, could pave the way for the first geological survey of Virginia. He also believed that few other scholars in the area could lead such an undertaking. "Yet when I recur to the still doubtful nature of my hopes of public employment in geology in Virginia," Rogers stated, "I almost decide for removal." He made it clear that his decision hinged almost entirely on the survey: "If I could be certain of obtaining the geological appointment this winter, I think that would decide me to remain here, unless, indeed, it could be combined with my duties in Philadelphia." When he succeeded in gaining support for the geological study, he finally resolved to stay in Virginia.[14]

If Rogers thought that the survey would free him from the stormy politics he'd found while at William and Mary, he was mistaken. Between 1835 and 1842 state leaders placed him in charge of a survey that drew him into repeated battles with the legislature. Southern scholars, like Rogers, faced mounting pressure to make their work relevant to southern causes such as the defense of slavery. As criticism from northern and international communities intensified, abstract or practical studies that failed to help defend the South met with opposition or even scorn. Similar sectionalism appeared within the state between regions east and west of Virginia's Appalachian mountain chain. Eastern politicians dominated state politics and represented the elite planter class. Their dominion crystallized after leaders from the western section of the state proposed political reforms that would have threatened slaveholder interests. The western population, far from being abolitionist, still included slaveholders. But their terrain—rugged, mountainous, and endowed with coal—led many of them to view agriculture as their past and industry as their future. For them slaves had less to do with economic growth in their region than state-led projects for the development of natural resources. Complicating matters, plantation owners regularly associated coal and other emerging industries with "Yankee" interests. Western-based political reformers understood this problem and how it had shaped political life in the state. "A large and decided majority of delegates and senators in the east," reported reformers, "had been insisted on as essential to the safety of the slave owners in that quarter and great efforts were made to alarm the holders of that species of property with the dangers that might arise from western influence in legislation." The challenge they faced was in wresting political power from eastern leaders, who viewed a gain for the west as a loss to the east.[15]

As each survey year passed, Rogers's study was drawn further into this struggle for power within the politics of slavery. At first state leaders interested in coal and state projects for resource development successfully backed the survey. Assembly member Joseph C. Cabell, a friend to Rogers's endeavor, had founded the Kanawha Canal Company for the purpose of advancing coal trade in the western part of Virginia. His support proved critical to the initial reception of the study. Kanawha County delegate George Summers also fought for Rogers. Reporting on the establishment of the survey, he declared, "The want of opposition to the bill today spoiled quite a scientific speech which I had mustered up for the occasion." The fate of Rogers's project in part depended on the extent to which eastern slaveholders would allow state projects that involved the promotion of northern-like industries in the western section of the state.[16]

The clash of interests between North and South, East and West, however, intervened in Rogers's program of research. His own conflicts with legislators pitted him against the dominant forces in the assembly. Although he attempted to strike a balance between surveying the natural resources of eastern and western regions, the project struggled to secure funds from year to year. Rogers lobbied an eastern-dominated legislature that often showed contempt for the work he was conducting. Referring to a set of scholarly reports on the survey, he fumed about the problem. "I fear there is trouble in store for me," he wrote, describing a turn of events. "By yesterday's papers I see that some resolutions offered . . . to aid the circulation and to enlarge the edition of the reports, met with great opposition, and that a long debate occurred, in which the merits of these documents were freely discussed." Here was the crux of the problem between Rogers and opponents of the state project. Rogers abhorred "the thought of a legislative body employing itself in venting spleen or exercising wit upon a paper of which but a very few of them have any adequate comprehension." The attack on the geological research, he exclaimed, "really fills me with indignation. It shows, too, that I have been mistaken in confiding in the good sense and good feeling of our legislature, and will destroy much of the satisfaction I have heretofore enjoyed in the prosecution of my tasks. . . . As it is, I am at the mercy of the ignorant or the illiberal."[17]

The political challenges persisted through the final years of the project. Rogers was appalled by "absurd speeches that have been made in and out of the house [of Delegates]." "An ignoramus who could not put two words correctly together," he declared, "made an attack in which he attempted to paint me as I addressed the house." The House member resorted to trivial tests of Rogers's geological knowledge with "a handful of stones before me, such as he could pick up anywhere in the roads in his county." Rogers found the experience humiliating, and it underscored for him the difficulty of attempting to advance an industry-related science in a state divided by the politics of slavery. Later he would face similar resistance to his ideas about scientific studies and educational reform. In the meantime he kept his eye on scientific developments in the North. The state surveys of New Jersey and Pennsylvania, led by his brother Henry, proceeded with fewer legislative travails and without the annual struggle for funds. At times the Pennsylvania legislature not only met Henry's budgetary requests but also exceeded them. For relief from the squabbles in Richmond, William turned to the national science community to stir an interest in the Virginia work. As the study progressed, the Academy of Natural Sciences of Philadelphia elected him correspondent, the American Philosophical Society offered him membership, and the National Institution for the Promotion of Science enrolled him as an affiliate. In the final years of the survey he also became a founding member of the Association of American Geologists and Naturalists.[18]

If Rogers found supporters in national organizations, they had little impact on the attitudes of Virginia's legislators. Calls for abandoning the multiyear survey appeared in the assembly of 1840 as the project faced intense scrutiny. Rogers lamented the "illiberal" character of life in the South and found few sources of support within southern higher education: "But how sad the contrast experienced here. . . . I feel that I am but half-alive here, and am more than ever resolved, when able, to quit the scene for one more congenial to my tastes and more likely to promote my happiness." In March 1841 one of Rogers's last sources of happiness dried up. The legislature sent him notice that funding for the survey would cease at the end of the year.[19]

Rogers requested an extension to complete the survey schedule year that began in April, but the petitions initially came to nothing. In fact, the opposition within the legislature reached such a feverish pitch that, as one observer mentioned, if supporters of the present bill "had not offered the form in which it is, one for an immediate repeal of your law would have been offered and carried." Rogers continued to argue in defense of the survey, but his efforts failed. He succeeded only in receiving compensation for the salaries of his assistants and other expenses to April 1842. The abrupt end to years of work left Rogers disillusioned. Unable to persuade Virginia's politicians to fund the publication of his final survey report, he turned to a circuit of presentations to disseminate his discoveries. Rogers delivered inaugural lectures at the Smithsonian Institution in Washington, D.C., spoke at scholarly meetings in Philadelphia, and gave lectures in Massachusetts about his work. Not surprisingly, when he compared the reception of his ideas among northern scientists to the political fallout in Virginia, he became hardened in his views. "Since my summer's rambles with Henry," he reflected about a vacation with his brother, "I have been unable to shut out the contrast between the region in which I live and the highly cultivated nature and society of glorious New England."[20]

While on the survey, Rogers accepted an offer that added to his tenure in the tumult, prolonging his final move to the Northeast. The professorship of natural philosophy became available at the University of Virginia the same year he began the geological work for the state. Rogers's name appeared second on the list of potential candidates. Joseph Henry, one of America's leading nineteenth-century scientists, received the first offer for the position. Henry, whose scientific experiments would later give rise to technologies such as the telegraph, was then at the College of New Jersey (later Princeton). Virginia's offer included housing, a reduced teaching load, resources and facilities for experiments, and a salary of approximately four thousand dollars. Compared to his twelve hundred—dollar annual wage at New Jersey, the offer proved tempting. But the problems associated with living in a slaveholding region, while not the deciding factor, seriously concerned him. "I do not like the idea," explained Henry, "of living in a slave state or going much farther south." Instead, he deferred to Rogers and sent a recommendation on his behalf. "Mr. William Rogers, of Virginia," stated the letter, "is well known as an ardent and successful cultivator of science. I am personally acquainted with him, and have a very high opinion of his talents and acquirements." "He is one of those," Henry suggested, "who, not content with retailing the untested opinions and discoveries of European philosophers,

endeavor to enlarge the boundaries of useful knowledge by experiments and observations of his own." So began Rogers's career at Jefferson's university.[21]

By the time Rogers arrived in Charlottesville in 1835, a tension existed between
the university's founding purpose and the changing southern context. The late
Thomas Jefferson's goal had been to establish a community of scholars in pursuit of
truth and led by an internationally recognized faculty. "This institution," he pro-
claimed, "will be based on the illimitable freedom of the human mind. For here we
are not afraid to follow truth wherever it may lead, not to tolerate error, so long as
reason is free to combat it." But long before Rogers's arrival the ambitious program
of tolerance and the pursuit of truth had begun to fade from the campus. John
Hartwell Cocke, who had assisted in the founding of the University of Virginia,
questioned the need to look abroad for intellectuals. "Do save us," he wrote to a close
associate of Jefferson, "from this inundation of foreigners, if it is possible." Students
also resisted the Jeffersonian ideal and acted on their animosity. A mob appeared outside the faculty residences late one evening during the institution's first fall term,
shouting, "Down with the European professors!" as they vandalized the foreigners'
property. Many of the original foreign-born scholars Jefferson had selected left the
campus and the South over the following years. Three out of four English faculty
members, for instance, "found the place not to their taste, and have left it," wrote
one observer. The fourth scholar openly expressed similar dissatisfaction and sought
a position in another region.[22]

The most controversial departure that left an impression on Rogers involved
mathematician James Joseph Sylvester, who had been elected to a professorship in
1841. After hearing that Virginia had selected an English Jew for the vacant position
in mathematics, some members of the southern community expressed grave dis-
pleasure. "This is the heaviest blow the University has ever received," decried the
Watchman of the South, a periodical of the Presbyterian Church. "The great body of
the people of this Commonwealth," continued the statement, "are by profession
Christians and not heathen, nor musselmen, nor Jews, nor Atheists, nor Infidels."
The author of the editorial sought to voice the concerns of many Virginians and reflected a widespread sentiment in the South. Although anti-Semitism certainly appeared elsewhere during the period, Sylvester's case was compounded by his English
origins, which doubly aggravated defenders of slaveholding traditions. "It is a historical fact," argued the editor, "that no Englishman has ever resided amongst us and
then written a book respecting us, without shewing his prejudice against us and his
ignorance of our peculiar Institutions." The fear centered not only on his religion
but also on the antislavery sentiment that he *might* bring to Charlottesville. Sylvester

received notice of the public outcry only shortly before setting sail across the Atlantic in November 1841.[23]

Rogers was stunned by the degree of intolerance in his midst. Although he might have expected some reaction from the community, given the previous incidents with foreign faculty and increasing sensitivity in the South over the slavery issue, he nevertheless found himself "mortified and provoked, too, at finding so much illiberality among a portion of the community here on the subject of religion, as displayed in the bigoted publications which appeared during the summer respecting the appointments of Sylvester and [Charles] Kraitzir." "Would you believe it," he remarked, "that a series of essays has been published condemning the Visitors for the appointment of a Jew and a Catholic?" Rather than provoke his southern colleagues, however, Rogers turned to his northern counterparts to express his concerns about the safety of the new professors and the future of higher learning in the South.[24]

Rogers soon discovered that Sylvester would last only six months. Without much support from the rest of the faculty, the mathematician faced anti-Semitic and antiforeign sentiment. Students shattered the windows of his residence, openly insulted him in his lectures, and physically assaulted him after class. In one altercation Sylvester defended himself from a student attacker, inflicting a wound that at first appeared life threatening. The faculty hastily recommended the professor leave immediately. Once the Englishman reached New York, Rogers supported Sylvester's chances of securing a position at Columbia College. He submitted a proposal to recognize the mathematician's departure as a voluntary resignation. Although the proposal passed a faculty committee, the Board of Visitors complicated the university's position, obfuscating the reason behind Sylvester's departure and appearing to dim his reputation. In the short term the imbroglio succeeded in hampering his career; in the long run Sylvester's flight from the South deprived the region of a scholar who later received prestigious academic awards, such as the British Copley Medal of the Royal Society, as well as appointments at the Johns Hopkins University and Cambridge University.[25]

As Rogers well knew, Sylvester's case was hardly an isolated problem. Student violence, whether over honor or the defense of slavery, had long frightened away experienced faculty and had caused rioting that at times required the use of the state's military force to bring calm. Rogers had already witnessed one of the most significant uprisings in the university's history, the Military Rebellion of 1836. In reviewing the incident, the chairman, John A. G. Davis, recalled how the "houses were attacked, the doors forced, and the blinds and windows broken. And there is reason to believe that not content with this, they contemplated proceeding to the desperate extremity of entering our houses for the purpose of attempting personal violence." Davis's report proved accurate for subsequent campus turmoil as well. After another series of riots a few years later, Rogers would write: "This morning I assisted in laying another of my colleagues in the grave. My kind friend . . . Davis died on Saturday evening of a wound received on the previous Thursday night! He was shot in cold blood while watching the movements of a student who, disguised and masked, was making riotous noises and firing a pistol on the lawn." The loss of Davis, in addition to the impact of Sylvester's departure and the Military Rebellion, left Rogers wondering if he should consider a position elsewhere.[26]

Before Rogers could follow this line of thought for very long, the faculty elected him chairman (equivalent to president). He agreed to lead the institution for the 1844–45 academic year, a period that was filled with continued student disturbances and legislative travail. His record of the term resembled those of previous chairmen, except that some crises began to affect members of his own family. One of Rogers's brothers, Robert Empie Rogers, had a few years earlier assumed the professorship of chemistry at the university. Robert soon came to realize how difficult southern academic life could be. Late one evening in 1845 students paraded and rioted near the faculty residences, causing a particular disruption at the chemist's property. After they had frightened his family, the professor hid outside the home until the students returned. As they approached again, Robert darted out from the shadows, captured one of the rowdies, dragged the flailing student over his shoulder into the house, and warned the rest of the troop that he was armed and willing to defend his family. Students gaped and returned to their rooms without further incident. After identifying the captured rowdy, Robert sent him to face minor academic disciplinary action the following day.[27]

While the episode ended quietly, William's role as chairman placed him at the center of what he feared would be a violent retaliation from the student body. For a brief period, he noted, the faculty had "acted a very manly part in arresting the excitement which this bold exploit of Robert created among the students." Yet the rioting returned, causing more disturbances to the faculty throughout the term. Although the violence ultimately resulted in the dismissal of a significant portion of the student body, an observer from Philadelphia wrote to the Rogers brothers, congratulating them "upon the triumph that you have obtained as a faculty over a set of lawless rioters. . . . The result, it is to be hoped, will ultimately redound to the benefit of the University, by evidence which it furnishes, that its statutes are not to be violated with impunity." "So far as I know," the letter assured them, "there is but one opinion entertained here on the subject and that is that you have acted wisely." Rogers may have agreed, but he continued to question whether it was wise for him to stay. Near the close of his term as chairman, he noted, "The annals of the college disturbances could hardly furnish another narrative as disgraceful to the character of the country as the history of this would be."[28]

Eclipsing the problems of student behavior, the Virginia legislature began to question the need for the university. Before stepping down as chairman, Rogers defended the campus from attempts by state lawmakers to abolish the institution's annual appropriation. The political struggle surfaced in part from publicized student disturbances that tarnished the reputation of the state's university. Legislators wanted to reevaluate the merits of continuing to support the troubled institution. Yet Rogers perceived the "illiberality" of the state legislature as the main impulse behind their motives. The state assembly, in December 1844, began an extensive audit of the university from its inception to the present, including "all and every appropriation of the University of Virginia at Charlottesville." Results from the audit produced a stir among house leaders and led to an inquiry "into the expediency of repealing the act of the assembly which authorizes an annuity of $15,000 per annum payable out of the revenue of the Literary fund to the University of Virginia at Charlottesville." The inquiry began a call for severing ties between the state and Jefferson's educational legacy.[29]

By January 1845 Rogers faced an increasingly contentious legislature. As university chairman, he prepared a lengthy defense for the institution to counter the "passions" rising from the House of Delegates. If he had thought that the hostilities came as a result of the publicized student disturbances, perhaps he would have tailored the report accordingly. Instead, Rogers focused his defense on the need to support the liberal ideals that went into the school's founding. In his report to the state assembly he argued for the benefits of the "intellectual culture" promoted by the university. "Through its well-trained alumni," he argued, the beneficial influences appear in "the methods and aims of academic teaching in many sections of the State." "In proof of this," he continued, "referring in the first place simply to the training of its own students in literature and science, whether professionally or with general objects, we would call attention to the extent and thoroughness of instruction which it offers, and to the system of intellectual culture it adopts." He emphasized the need for a strong academic presence in the state and called on legislators to reconsider the proposal to abolish the annual appropriation.[30]

Rogers buttressed his arguments with four subsequent points. First, he reminded state leaders of the institution's contributions to the nation in advancing a new system of higher learning. The distinctive features supported by the founder included the elective system and extensive use of the lecture method, as opposed to recitation. Rogers defended the elective system as having a "wise regard to the practical wants of society." It addressed the need to provide a varied but rigorous course of study to students with varied pursuits. On the delivery of such instruction, Rogers noted the positive influence for professors and students of the lecture method, claiming that with this mode of teaching the professor could make lasting impressions on the minds of youth and elicit enthusiasm, while students could exercise higher faculties of listening and criticism. Second, he reminded them of the university's policy against honorary degrees, a policy that aimed to protect the institution's credibility. The common practice at most colleges of awarding advanced diplomas to students without additional studies had never become a practice in Charlottesville. Third, the chairman felt compelled to address the legislature's concerns over faculty salaries. He pointed to the pay that professors received at comparable institutions of the South and Northeast, describing the disparity as problematic. In fact, he argued, the "compensation formerly given, has proved, as is well known, insufficient in some instances to secure the services of distinguished scholars invited to its halls, and has not prevented the resignation of many professors who had for a time filled its stations with undenied success." After noting the loss of professors, the report turned to the fourth and final point: how the annual appropriation in question compared to institutions in almost every region of the country. His assessment concluded that the funds appeared pitiably small. Nevertheless, Rogers warned that, if the university had to resort to private support, "the general interests of the community, as affected by the operations of the institution, would be either wholly neglected or but partially secured."[31]

The report enjoyed moderate success in the assembly, but several state leaders continued to challenge the appropriation. Pressure to disband the institution increased, stirring proposals for a merger between the state's Military Institute and the university. One informant recommended to Rogers that merging was perhaps the only hope for a recovery, particularly under the "frequent abuse which has been of late unjustly heaped upon her [the university], and the strong feeling of hostility manifested in the present legislature." Rogers, however, staunchly opposed the idea. "We are far from being discouraged by the result," he responded, "and are still strongly hopeful of steady and increased success, notwithstanding the ungenerous enmity of those who, from prejudice or ignorance, are laboring for our overthrow." The legislative session ended, and Rogers's optimism proved warranted, for the appropriation survived. While pleased with the success, he had turned his sights northward "to look . . . for some other and more tranquil home."[32]

In March 1848 Rogers submitted a letter of resignation to the University of Virginia that surprised many on the campus. Their surprise revealed that few at the institution understood his concerns. Although appeals from students and faculty led him to postpone his resignation for five years, Rogers had long resolved to leave the region.[33]

Between the first resignation and his final departure, Rogers married a northerner, Emma Savage, whom he had met in the mid-1840s during a research trip to New England's White Mountains. Emma and her family represented the reformist spirit Rogers longed to see in the South. Among Boston Brahmins they were known for their "hatred of intolerance and bigotry" and for helping to initiate the state's first public system of primary education. Emma joined Rogers during his final years in Virginia, but in the spring of 1853 they left for Massachusetts. Rogers had little more than an idea to look forward to in the North. "Ever since I have known something of the knowledge-seeking spirit, and the intellectual capabilities of the community in and around Boston," he mused, "I have felt persuaded that of all places in the world it was the most certain to derive the highest benefits from a Polytechnic Institution." Beyond this he uprooted his family without expectations for building a second career, exchanging "certainty for uncertainty," as one family member put it.[34]

During the many years leading up to this decision, Rogers had stayed in the region in part for the professorship in natural philosophy at the University of Virginia. Housed within the institution's Jeffersonian traditions, the faculty position, whatever the conditions of campus life, gave Rogers a source of recognition within the broader scientific community. In addition, his role as chairman lengthened his tenure in Charlottesville. The opportunity to lead an institution would have appealed to his interest in higher educational reform. But Rogers's primary reason for staying in Virginia was his scientific research. His studies off-campus literally kept him on the state's soil, in its springs, and on its highest summits.[35]

CHAPTER THREE

From Soils to Species

D ESPITE THE CHALLENGES he faced in the South, Rogers established a career in geology and natural philosophy. He did so with a worldview organized around the useful arts. By adopting this emphasis, Rogers sidestepped the two most common models available to early-nineteenth-century scientists. On the one hand, Baconians, proponents of the first model, generally followed Francis Bacon's inductive approach to science, which valued fact collecting and discouraged theorizing or generalization. The kind of Baconianism that arrived in the United States came with a Scottish accent. Common-sense philosophers of the Scottish tradition influenced a spirit of science in America that was based on the collection and classification of facts—facts that they hoped would one day form the foundation for grand laws of nature. Humboldtians, on the other hand, found Alexander von Humboldt's approach to science more compelling. Employing a model based on his approach, scientists questioned the value of an isolated fact. As early as 1805, Humboldt questioned the futile efforts of "travelling naturalists" and their concern "exclusively with the descriptive science and collecting." This German tradition emphasized the interrelationships of all observable phenomena, an approach Humboldt dubbed as "terrestrial physics." The goal of terrestrial physics was to uncover "the great and constant laws of nature" through comprehensive analyses involving time, pressure, altitude, magnetic and gravitational forces, chemical composition, atmospheric tests, and so forth. A rock found in the Andes mountain range of South America, for terrestrial physicists, would only gain meaning by analyzing as many physical factors surrounding the rock as possible. In isolation it meant little.[1]

Neither Baconianism nor Humboldtianism adequately defined Rogers's approach to science based on the useful arts. He agreed with Baconians that the storehouse of American scientific knowledge was inadequately filled. He saw the need for continued fact collecting to make any reasonable generalizations. Not enough was known, for example, about American geology during the early part of his career to allow for a simple consensus on the names used in stratigraphy, the study of the earth's layers. Yet he disagreed with the Baconian resistance to theory, placing him among Humboldtians. Rogers understood the need for fact collecting as part of the theory-building process. The Humboldtian pursuit of "the great and constant laws of nature," however, appeared at times on the margins of his program in the useful arts. Certainly, America's quest for practical knowledge informed his ideals, modifying any Scottish or Germanic influences. Of this practical quest French author Alexis de Tocqueville, in *Democracy in America,* described Rogers's scientific milieu as "filled with discoveries immediately applicable to productive industry." Tocqueville ascribed America's emphasis on utilitarianism, unlike European science's emphasis on theory, to the nation's "democratic, enlightened, and free" character. The aristocracies of the Old World, he argued, produced a vastly different culture of science than that emerging in the democracy of the New World. For the French author politics shaped science.[2]

In the mid-1830s, while Tocqueville was penning his comparisons between scientific cultures, Rogers had begun to establish his vision of the useful arts in geology and natural philosophy. The idea encompassed Rogers's mission to strike a balance in advancing both practical and theoretical knowledge. His aspirations in science extended beyond merely developing technologies or rationalizing the world around him. As a geologist and natural philosopher, he followed American traditions and practiced European methods. While much of his work aimed to provide useful knowledge to students and colleagues as well as farmers and politicians, another portion of his research program spoke to his counterparts in Europe. These divergent interests—American and European, practical and theoretical, geological and natural philosophical—characterized the scope of Rogers's scientific career. Yet, like almost all scientists of any era, Rogers did not work alone. His interests, discoveries, and writings developed through collaborations with many others who became part of an informal network of scientists. This informal network included family, friends, colleagues, and assistants who shared ideas or assisted in discoveries that supported Rogers as he undertook sometimes grand projects in the world of antebellum American science.

By the time Rogers took seriously the geologist's calling, there were few in the United States who had made a living from this branch of science. Most earth scientists of the antebellum period came avocationally from the ranks of trained physicians. The first so-called father of American geology was William Maclure, a Scottish immigrant with European training in the sciences. In 1809 he published the results of a geological tour of the eastern United States. Maclure's Baconian-styled work called on earth scientists to collect facts rather than debate theory. In the 1820s Amos Eaton, a former student of Benjamin Silliman, replaced Maclure as a central figure in American geology. Eaton's credentials included the *Index to the Geology of the Northern States* and an extensive five-year geological survey of Albany and the Erie Canal published in 1824. In the tradition of Maclure, he focused on the practical application of scientific knowledge, emphasizing the utility of identifying deposits of coal for the emerging industrial boom that was then occurring in the Northeast. A new band of geologists who made careers out of the field began to appear in Philadelphia in the 1830s. Affiliated with the Franklin Institute, the group included Alexander Dallas Bache, Samuel George Morton, and George Featherstonhaugh.[3]

Inspired by his family and a widening circle of geologists in America, Rogers began looking for an outlet to publish his own initial findings in the field. The most important journal for Rogers's early work was Edmund Ruffin's *Farmer's Register.* Ruffin, known best for his fire-eating southern nationalism, tried for many years during the antebellum period to give the southern agricultural community a scientific voice and resource. Ruffin's concern over the rapid depletion of top soil minerals as a result of poor farming practices first motivated him to publish the journal. Through the *Register* Rogers contributed to the reform efforts with essays on fertilizers, soil analysis, and scientific techniques. Both Ruffin and Rogers believed in bringing science to farmers of the southern states. The main difference between the two centered on their scientific training; Rogers had been raised for a life in science, while Ruffin had received almost no training at all. To bolster the "science" of Ruffin's agricultural movement, Rogers sent several articles to the Richmond publisher.[4]

Rogers's essays for the *Register* centered on marl, a soil type also known as green sand. Many Virginians came to view marl as a potential answer to the state's troubled agricultural economy. As a result of the panic of 1819, prices for the state's crops had plummeted. Planters answered the price drop by overworking their lands with reckless farm practices that depleted the soil of its productive elements. Convinced that the damage to the Virginia fields was irreparable, many planters moved westward to untilled lands. Between 1817 and 1829 Virginia farmland values declined by more than half, and over the course of the following decades the state lost nearly four hundred thousand inhabitants. By the time Rogers and others began to study the efficacy of marl as a manure (soil additive) in the 1830s, Virginians were searching for a way to stem the state's steady decline. The social and economic conditions of Virginia deeply influenced Rogers's early scientific agenda. Thus, while his research on marl emphasized practical geology, it was not to the exclusion of basic scientific knowledge.[5]

The practical studies that Rogers conducted on marl included chemical analyses and the development of technological innovations. Chemical analysis formed an integral component of Rogers's geological work on the green sand. To talk with farmers and analyze their soils, he visited farms in New Jersey where marl had been used. He made dozens of analyses from the soil samples he collected, estimating the percentages of "green particles" they contained. The marl analyses were part of Rogers's efforts to cast "some light upon its agency when applied to the soil." He described methods for separating green sand from other soils and discussed the shell and fossil content his studies revealed. Rogers recognized that he worked in concert with other scientists, and not alone, when he stated that "these results agree very closely with the determination of [Pierre] Berthier of France, and [Adam] Seybert of Philadelphia. The former operated upon the green sand of Europe, the latter upon that of Philadelphia." His analyses also sought to dispel myths about the organic and inorganic composition of other manures such as the oyster shell. "This is a question of some interest," he wrote, "in the agricultural application of the shell, since the form in which it may be most usefully employed as a manure, will depend upon the quantity of animal matter it contains." In his research Rogers pulverized, heated, weighed, and acid tested the shells to determine the shell's composition. He found that, contrary to farm lore, "no appreciable advantage can be expected in applying it as a manure from the minute portion of animal matter which it has been shown to contain." Rather, other elements present in the shell "should claim the attention of the agriculturist" for improving the fertility of soils.[6]

The chemical analyses published by the *Register* spawned a desire among southern farmers to have their soils analyzed, inspiring Rogers to construct new instruments for his practical work on marl. From his teaching experience he understood the value of technologies for science instruction and knew how to construct such laboratory equipment. When readers of Ruffin's journal began to overwhelm him with packages of soil samples for him to analyze, Rogers applied his technical knowledge to develop new instruments to handle the requests.[7]

He had started analyzing marl by using "the instrument of Rose" but found several flaws with the apparatus. For research purposes the system was "cumbrous and difficult of management." Over the course of his soil experiments the Rose instrument had been prone to accidents and difficult to regulate. These and other deficiencies had rendered "an exact estimation" of results from experiments nearly "impossible." Rather than continue with imprecise measurements, Rogers constructed a more accurate and easy-to-use alternative. He explained to his readers how to construct replicas of his instrument and how to use it for marl analysis. He gave no name to the innovation other than an "Apparatus for Analyzing Marl and the Carbonates in General." A second instrument, one he called a "Self-Filling Syphon for Chemical Analysis," had similar origins.[8]

When it came to abstract work on marl, Rogers often played the role of advocate for basic research. His essays discussed the need to understand "geological laws" for practical purposes. He wrote lengthy accounts of geological observations he had made of New Jersey land forms to assist Virginia farmers in finding similar forms in their state. Through such a comparison he encouraged readers to find layers of earth or strata rich in marl. Rogers used marl as an incentive for farmers to learn basic geological principles. "Such facts," he argued, "are frequently invested with a *practical interest,* by the aids which they furnish to other and more important discoveries." He introduced readers of the *Register* to stratigraphy, the study of the earth's layers and land forms, acquainting farmers and educators with the geological debates of the period to encourage accuracy in identifying geological formations. By doing so, Rogers hoped to widen the circle of correspondents who might help expand knowledge of Virginia's geology. At the same time, he wanted agriculturists to make reasonable conclusions about the location and extent of marl beds for their own practical purposes. These conclusions were not irrelevant to basic science, for Rogers was engaged in the debates of the era over whether strata should be identified by its mineral or fossil content. Amos Eaton argued for using minerals, the traditional way, while English geologists began adopting new methods based on fossil or hybrid analyses. Rogers sided with the English, especially when he declared that strata should be identified with "fossils, whether shells, bones, or vegetable remains, which the strata contain—a procedure to which [scientists] have been led by the whole tenor of modern developments in geology." As a basic research advocate, Rogers attempted to reconcile his desire for diffusing modern geological thought with the farmer's desire for practical knowledge.[9]

He also made use of the popular interest in marl to highlight some of his own basic discoveries. The precision of his new laboratory technologies, for example, allowed him to revise the work of English chemist Charles Hatchell. "Some of his details," Rogers explained, "especially in regard to the oyster shell, seems to have fallen into error." With the help of more precise equipment Rogers responded to European debates, aligning his work with such German scientists as Christian Friedrich Bucholz and his student Rudolph Brandes. Both men had been recognized on the Continent for their work on separating chemical elements such as magnesium and calcium. Rogers combined his understanding of their work with his technological and scientific interests to produce new knowledge. Outside the laboratory Rogers made new observations about Virginia's geology. He investigated "the existence of a lower tertiary deposit throughout an extensive district of Eastern Virginia," which he described as the layer of earth closest to the surface, one that contained a mixture of identifiable and unidentifiable fossils. Secondary deposits, such as one Rogers had examined in New Jersey, existed beneath the tertiary layer and contained mostly unrecognizable fossils. The oldest and deepest layer, the primary deposit, contained no recognizable remains. Rogers's research on marl had led him to uncover "hitherto unknown" land forms and deposits in the state.[10]

~~

During these early years of soil research Rogers's passion for the advancement and diffusion of knowledge crystallized. His concern for the intersection of the practical, the theoretical, and the technological would become hallmarks of his plan for MIT. To advocate such a plan, however, at least successfully within the science community, one needed scientific credibility. Collecting soil samples is one thing; gaining name recognition is another. Rogers was ambitious, though, and wanted to make his mark in the field of science, not knowing that doing so would serve him well later with his educational reform efforts. For those with ambition during the antebellum period, large-scale research projects offered the fastest way to achieve scientific credibility and recognition. In this era, at a time when graduate school in geology was not an option, directing a survey was the next best thing. Amos Eaton had made a name for himself through his New York geological work. Alexander Dallas Bache would later become a star as superintendent of the United States Coast Survey. Rogers wanted a project of his own. And he found one possibility when Virginians began to clamor about wanting a state geological survey.

Rogers still had his hands in the dirt, collecting samples, when the first push for such a survey came from outside Virginia. Geologists in Pennsylvania had their eye on the state as a site for a comprehensive, state-sponsored survey. George William Featherstonhaugh and Peter Browne, both members of the Geological Society of Pennsylvania, had worked for years on scientific and political fronts to justify the expense. Featherstonhaugh, a geologist of English origins and scientific experience, used his expertise to make a case for the value of studying Virginia's natural resources. In particular, he wrote on the area's mineral deposits and petitioned such state leaders as James Madison and Joseph C. Cabell to back further research. Peter Browne expressed his interest in conducting a statewide study in a letter to Governor John Floyd in 1833, recommending nothing less than "a topographical, geological, mineralogical and orgetological survey of Virginia."[11]

Of the two Browne came closest to establishing himself in the desired seat of state geologist. But he made a few missteps that Rogers managed to avoid, thanks to his experience as a citizen of Virginia. Browne invoked arguments that, while effective in the North, were less persuasive in the South. He argued principally for the exploration of the state's coal resources for industrial purposes, failing to realize that northern interests in coal were not as strongly reflected in the agrarian South. Browne also argued that the survey would produce advances in practical and theoretical knowledge. "It is the duty of states," he proposed, "to furnish their quota to the general stock of information." Yet he overlooked the comparative indifference toward such goals in Virginia, one that Rogers had observed as a science professor in the region. Browne's final argument invoked theology for support. Drawing on religious sentiment for such a project, however, might have had greater force in Massachusetts or Pennsylvania rather than Virginia, where Thomas Jefferson had founded one of the country's first secular universities. Thus, when Browne suggested that the survey would invite "the mind of men to reflection, and his hands to industry, and displaying at every step the wisdom and beneficence of the great Creator," he did not expect to sour his chances.[12]

Rogers, of course, had the advantage of knowing the state and its potential interest in a survey. When he came before the legislators and the governor, his arguments reflected this advantage. Agriculturists, for instance, topped the list of beneficiaries throughout his proposal. He understood that the dominant political forces in Virginia looked to marl, not coal, for a solution to their state's problems. Rogers assured them that his investigation "would be applicable to all sections of the state, and which would undoubtedly contribute to the general benefit of our agriculture." To this end he promised a "systematic analysis of all the important varieties of soil within the state." Mineral wealth was not far behind on the list of benefits. But rather than vaguely refer to metallic ores, as one section of Browne's petition had, Rogers identified a resource that was sure to capture the attention of state leaders: gold. His survey would outline the "imperfectly traced" gold deposits throughout Virginia. He hardly had to mention that "to trace out the gold region entirely through the state, would be an important and useful work." The most clever appeal, however, was not based on agricultural or economic motives but on the Virginian psyche itself. If there was one sensibility that nearly all Virginians shared, he concluded, it was pride of country and state. Thus, Rogers spared no opportunity in tapping the nationalistic spirit when he alluded to the geological work by the governments of France, Sweden, Russia, Germany, Italy, and Great Britain. These countries had sponsored "detailed inquiries into the geological features and mineral resources of their respective domains." Tapping state pride aimed for a similar effect. North and South Carolina founded the first American state surveys in the 1820s, followed by Massachusetts, Maryland, Tennessee, and Connecticut in the 1830s. Woe to the Old Dominion, implied Rogers, if it failed to follow "the wise example" of these states.[13]

The persuasive rhetoric, coupled with his practical work on Virginia's geology, distinguished Rogers from his competitors in this struggle over the survey. When he made a final presentation to the state assembly, he sensed that his approach had carried the day: "I marched into the hall of delegates" and advocated "the cause of geology, developing a few of its most important truths, and displaying the benefits which it proffered to Virginia." For over an hour Rogers held the attention of the House, closing his address to "loud words of approbation." "Friends say," he rejoiced, "that the legislature will authorize a reconnaissance this year, and of course I shall have the management of it." His oration inspired the act establishing the reconnaissance for the survey on March 6, 1835.[14]

As director of the study, Rogers stretched the funding from this act, and one passed the following year, across seven years of geological research. This period of research produced the first systematic survey of Virginia, consisting of seven annual reports. Rogers understood his main goal as preparing a comprehensive review of the geological and mineralogical features of the state. The questions he took with him into the study were therefore largely practical, while the methods for answering them were grounded in theory; thus, he combined Baconian and Humboldtian models. He sought "local information of a useful nature . . . consistent with the time which could be devoted to investigation of a special character." By "useful information" Rogers meant the identification of precious or productive natural resources located within the state's borders. He aimed to provide the exact locations of areas likely to prove valuable to Virginians, and, like other geologists of the period, he relied on two common methods to achieve his goal: field observations and chemical analyses. The survey, published in the annual reports, used descriptive language based on field observations. Detailed notes taken in the field formed the basis for a systematic description of coastlines, mountain ranges, river patterns, and other striking features in the landscape. In the laboratory specimens collected on field excursions were subjected to chemical analyses. Hundreds of tests revealed the interior composition of those natural resources described by their exterior features in the observation notes.[15]

The vast and varied terrain of Virginia led Rogers to believe that the survey was "the largest area ever subjected to systematic examination in any part of the world." Whether or not this was an exaggerated claim, he wanted to make his mark in geology through the ambitious survey.[16]

The mark it left was in the useful arts. Agriculturists, for example, were among the best served by the study. For farmers, who steadily sought to locate minerals with which to replenish their soil, Rogers offered a general description of the landscape as well as a detailed analysis of specimen samples. When he came across such natural exposures as cliffs, he would observe their structures: the thickness of the layers of earth exposed; the arrangement of beds of clay, sand, and soil; the color and texture of the various earths. His work brought to light the browns, yellows, blues, greens, and grays of the soils and their fine or course, soft or hard, textures. He would identify the height of land forms, the distribution of comparable forms across the region and state, and then determine the exact location of each. When it came to content, Rogers looked for the obvious and the less obvious; shells and fossils, for example, stood out as prominent features in the soils of the marl region. He paid particular attention to the contents of the blue marl and the yellow marl, identifying the organic and inorganic matter in each. The most important chemical test Rogers conducted on the marls determined the percentage of carbonate of lime contained therein. Soils with high content of carbonate of lime generally proved to be more fertile, a lesser-known piece of agricultural advice at the time. In the descriptive language and quantitative analyses of the reports, the survey acquainted agriculturists with their surroundings from an external and internal perspective. Rogers provided a means for them to make informed decisions about the soils they used to fertilize their crop fields.[17]

Miners, builders, and architects also benefited from the survey. In it Rogers pointed to valuable deposits available for exploration. The survey described the boundaries and thicknesses of coal beds as well as gold, copper, and iron deposits scattered throughout the state. Laboratory tests measured the heating capacity and the composition of combustible material and the purity of precious metals. Builders and architects who worked with sandstone could make use of Rogers's depiction of sandstones of various coloring, including dark-green, light-gray, blue, purple, and greenish-grays. In the mountain region he sought to determine "the true relation of the rocks and minerals of this portion of the state." In doing so, he argued for the necessity of using science as the foundation for determining useful quarries and similar rock formations. He considered the importance of identifying varieties of rocks "according to the names by which in scientific language they are respectively known." Common names, he explained, had muddled the terminology of earthen materials, complicating the task for practitioners and theorists alike. As a result, cultivators of science had carelessly confused "granite and gneiss with sandstone, mica slate with soapstone, and ores of iron with those of silver, lead or gold." The so-called technicalities of naming objects had relevance not only for fieldworkers but also for those in "the workshop, the warehouse, the plantation or the mine." In each scenario adherence to scientific terminology, rather than common names, provided a means of communicating with fellow scientists, comparing or contrasting objects from distant locations, and distinguishing materials of a like nature.[18]

The names and composition of rocks, he claimed, were central to the tasks of the state's builders and engineers. Rogers directed the survey, for example, to determine the structure of the mountain formations for internal improvement purposes; the reports noted thickness, contents, textures, and depths to determine the best means of developing the region. "We may derive a suggestion," he posited, "of some importance in connection with plans of internal improvement projected in the state, which is, that the dense and impracticable character of many of the rocks . . . forbids any attempt at tunnelling." If Virginia planned to extend roads and railways throughout the mountains, Rogers's survey promised information of a useful nature for the enlightened management of such efforts.[19]

Every section of the survey, whether directed to farmers, miners, builders, or others, emerged out of collaborations of one form or another. Rogers received the support of nearly a dozen field and laboratory assistants, ranging from novice to expert. If naturalist manuals of the era are any indication, Rogers fitted himself and his assistants with the thickest-soled boots they could find. As they traveled together by foot and by horse and buggy during the seven years it took to complete the study, they carried with them at least the basic tools of the geologist's trade. A light hammer would be needed for small rocks, separating fossils and shells, working with crevices. A heavier hammer provided them with the means to break open larger rocks and to extract specimens from hard formations. A third hammer available to them was a small pickaxe, or "platypus pick," named after the pointed and flat ends. Geologists commonly used this tool on excursions involving clay and similar deposits.[20]

Using these tools, Rogers and his assistants collected rock samples, often a few inches wide and an inch thick. Mineral and fossil samples ranged in size and variation, crystals being particularly prized findings. For sands, clays, and other soil materials they used vials made of glass or wooden boxes. For each specimen collected on the excursions, labels and field notes described the exact locations of the extracted samples. Notetakers also entered observations about the formations from which the samples were taken; such observations included records on the thicknesses and contents of the formation and surrounding formations, the predominant shells and fossils in the area, the position and direction (horizontal, vertical, or inclined) of related land forms. For outings to the state's highest peaks Rogers brought along a few extras. "Observations for determining the altitudes of the principal ridges and escarpments of the region we were exploring," he noted, "were carried during the season by means of barometers and *altitude* thermometers." The thermometers, in particular, he added, had provided "accuracy of the heights computed" to form topographical estimates for mapping the region.[21]

Because of a lack of support from the legislature, Rogers used personal funds to establish an adequate laboratory for the survey. The laboratory housed the apparatus for analyzing carbonates, as described in his earlier research, among weights, scales, and other chemical or mechanical equipment necessary for mineralogical analyses. Also in the laboratory, or nearby, Rogers assembled a specimen cabinet for museum and educational purposes. The cabinet aimed to provide a mineralogical portrait of the state's geological features. For traditional geologists these tools and methods were used for determining the history of the landscape. Rogers applied the same tools and methods for the useful arts.[22]

The assistants Rogers hired often needed training in the field and in the laboratory, a problem that partly inspired his later proposals for a technological institute. For a time he followed his brother Henry's advice, who was then working on the surveys of Pennsylvania and New Jersey, that "under no circumstances ought the state to look to us for detailed work; that is to come from the assistants." But William quickly realized that his absence in the field created hardships, confusion, and frustration among his assistants. While Henry became embroiled in bitter debates with assistants over credit for major discoveries, William's problems were of a more practical nature. He received notes of poor quality from his aides scattered across Virginia and remarked that his occasional visits to the geological sites proved more useful "than months of comparatively blind labor of . . . assistants." Making matters worse, William had difficulty retaining competent field researchers. Henry attempted to send at least two qualified surveyors to Virginia, but apparently both of them declined. With few exceptions William employed assistants who either left after a short period or lacked basic qualifications for the task. In deciphering notes of uneven quality, losing workers, and spending time training new ones, Rogers faced a daunting obstacle to the survey's progress. The nature of doing fieldwork created challenges for those researchers toiling under adverse conditions. George W. Boyd, one assistant in Virginia, lost his life while working on one of the more difficult phases of the survey. Boyd and other field researchers who risked their lives on surveys across the nation reminded their colleagues of the serious hazards inherent in their line of work.[23]

Despite mishaps and tragedy, the most productive collaboration between Rogers and his assistants was with his brother Henry, who aided William when not consumed with survey work of his own. William and Henry approached the Appalachian Mountain chain in a way that revealed a persistent focus on the useful arts, particularly in relation to debates over basic geological knowledge. They devised a utilitarian method for identifying layers of rock formations present in the mountain chain in an era when American geologists hotly debated coding schemes. During the 1830s and 1840s nomenclature in geology had not yet passed a formative stage. No single system had gained dominance in America.[24]

The Rogers brothers proposed a numbering system as the most useful and practicable, a system designed to counter the proliferation of local names to describe similar formations located in different states and countries. James Hall of the New York survey, however, preferred to organize American stratigraphy according to the local names he and his assistants employed in New York. While the brothers focused on utility, Hall aimed at priority. Because William and Henry did not effectively publicize their numbering system to the American geological community, they left no lasting imprint on stratigraphic thought. Yet the emphasis on numbers, as opposed to names, reflected Rogers's useful arts bent. It provided a means for a majority of naturalists, agriculturists, miners, and others to focus on the character of a formation in relation to the number associated with it, rather than to local names of value to only a few. Rogers applied a similar system with letters to define the stratigraphy of agricultural soils. Rather than fix names to the layers, the survey assigned a letter system to what he called a "geological column." The column illustrated a cross-section of the earth's surface from the topsoil to the soft and hard rock below. With the column and lettering system farmers could estimate relative distances between one layer and another to reach the desired marl or soils or clays that could then be used as fertilizer. Both schemes, for the Appalachian and the marls regions, provided flexibility during a period of competing coding systems in geology. The Rogers scheme encouraged nonspecialists to draw useful insights from geological research while avoiding the multiplicity of names. If the New York system, a European system, or any later system became the standard, they thought, the numbers and letters could easily be replaced with the established code.[25]

Rogers maintained an emphasis on practical goals and scientific methods over the course of the survey, with the exception of a second collaboration with Henry. In a paper read before the Association of American Geologists and Naturalists meeting of April 1842, William and Henry presented a new theory of mountain formation. The presentation centered on the work conducted by both brothers on their respective state surveys. William had collected observations about the Appalachian chain over the previous seven years in Virginia, while Henry studied the same mountain system during his geological work in New Jersey and Pennsylvania. From their findings the brothers developed a general theory that they believed promised to explain all mountain-building phenomena.[26]

Before the nineteenth century catastrophism dominated much of the discourse over mountain building and the general formation of the earth's crust. Proponents of catastrophism, while not a monolithic group, generally assumed that unusually violent upheavals—floods, earthquakes, volcanic eruptions, or other agents of change—occurred during the forming of the globe. German lecturer and theorist Abraham Werner became a leading representative of a catastrophism that argued for the "deluge" theory. Through a process of evaporation, he stated, land formations have emerged out of a catastrophic, "born from chaos" flooding of the earth. The theory proved attractive to scientists desiring to reconcile the emerging study of geology with the biblical understanding of a Noachian flood. Another reason for its popular adoption stemmed from Werner's charismatic lectures and broadly strewn disciples. Yet by the early nineteenth century a new theory began to displace the old. Uniformitarianism, as articulated by Werner's Scottish rival James Hutton, rejected the notion of unseen geological processes and proposed a more gradual development of the earth's surface. His theory declared that the "present is the key to the past." Both theories, however, lacked a substantive empirical grounding, and debates between Huttonians and Wernerians continued until British geologist Charles Lyell appeared with his *Principles of Geology* in the 1830s. Lyell popularized uniformitarianism in the English-speaking world and helped launch the field of modern geology based on observation. Continental geologists, meanwhile, continued to elaborate on catastrophic theories because many of them believed Lyell had unsatisfactorily explained the formation of mountain ranges. In large measure the Rogers brothers hoped to close the gap in the geological research, a corrective that would attract attention from European scientists.[27]

At the time William and Henry proposed their theory, geologists generally held one of two conceptions for explaining the Appalachians and other mountain systems. First, the chain could have resulted from a vertical force. Through violent, "paroxysmal upheavals" the formations could have come into being. Second, they posited, horizontal forces could have also caused the folds in the earth that led to the chain formation. As a precursor to the now accepted plate tectonic theory, some postulated that "a horizontal or tangential pressure" had likely contributed to the mountain chain. In this context William and Henry offered a third possibility: a combination of the two leading views. The Appalachians and all other such systems, the brothers theorized, occurred by both horizontal and vertical forces. To account for both they conceived of a "wave-like oscillation, and a tangential or horizontal pressure. . . . This oblique inflection of the strata will, we confidently believe, be found to prevail as the regular form of all" similar formations "in every part of the world." Drawing from their observations of the Appalachian range, they generalized that "all great *paroxysmal actions,* from the earliest epochs, to the present time, have been accompanied by a wave-like motion of the earth's crust." The brothers also believed that the same processes affected earthquake motion, glacial drift, and coal formation.[28]

The theory, as presented and subsequently published, received three different reactions. In the United States, a setting of practice rather than theory, the Rogers wave concept attracted little immediate attention. Few, if any, American geology textbooks of the decade mentioned or referred to the theory. British scholars, however, found the notion untenable but debated its merits in science forums. Critics in England, such as Adam Sedgwick and Henry De la Beche, either believed existing catastrophic theories (i.e., vertical or tangential motion) were sufficient to explain mountain chain formation or sided with a growing consensus that uniformitarianism would replace catastrophism. A few scholars in Great Britain and a much larger pool of French geologists responded favorably to the Rogers theory. Supporters argued that the idea had extensive grounding in the detailed surveys of three states. More important to French researchers, the wave idea reinforced their own contentions about catastrophism against the encroaching Lyellism from England. Henry, on a return visit to Europe in the late 1840s, declared France a more hospitable place for the wave theory than America or Great Britain. He told William that their theory was "well-known in Paris by the geologists" and had met with "general approval," despite the fact that at home they faced "bitter opposition." The Rogers "doctrine of flexures [the wave theory] produced by an undulation of the crust," he believed, would "meet the prompt reception by the French geologists, even while many of the English may hesitate." In the end the Rogers brothers were more successful in spurring theoretical debates in geology rather than advancing any particular "doctrine."[29]

The Rogers theory, generated from the geological surveys, drew from a parallel branch in the development of William's scientific career. During the period in which he directed the Virginia survey, Rogers had also been engaged in research in natural philosophy, a conglomerate of fields mostly to do with physics. The work on mountain formation theory unveiled only part of his interest in dynamics as it related to geological phenomena. The field of natural philosophy attracted Rogers to other areas such as engineering and physics, or what he called mechanical philosophy. As with geology, his interest in natural philosophy centered on the ideal of the useful arts, encompassing practical as well as theoretical emphases. Unlike his geological work, however, he produced no equivalent to a sustained, survey-like study for natural philosophy. Instead, Rogers published numerous short articles, introductory books, and science-related addresses that revealed his research interests.[30]

An article on the laws of the "Elementary Voltaic Battery" illustrates well the practical and theoretical interests that stretched across his other publications. Co-authored with his brother Henry in 1835, the article records a series of experiments that led to the development of a new research technology as well as new challenges to accepted laws in natural philosophy. The technologies used for this study involved a "voltaic battery" and a galvanometer. The brothers tested the conductivity of different metals in the battery acid solution and soon found that improved means of measurement required a new research technology. To measure more accurately and record the activity in conductivity, they developed the "torsion galvanometer." With a torsion key, the galvanometer could be preset to a specific range. As the spikes in activity occurred, the needle could be moved to a greater or lesser number on the index until no movement was measured.[31]

The new instrument allowed William and Henry to question the validity of accepted laws in "electrical science." Through a series of experiments involving a variety of metals, weights, and length of time exposed to the acid, they concluded that established theories were based on inaccurate data. "In most of the experiments hitherto performed by others on the elementary battery," they noted, "attention has not been directed to separating the momentary from the permanent effects, nor indeed to estimate the momentary effects themselves with any thing like precision." Moreover, they stated that "the proper permanent effects have not, as far as we are informed, been the subject of any observation at all." Armed with new, more accurate results, William and Henry challenged the work of British chemist William Hyde Wollaston on the subject of ratios compared: "Wollaston's plan of enclosing the zinc on both sides with copper, is really less advantageous than allowing the copper to extend to a double length on one side of the zinc plate. The fact of the copper being presented to both zinc surfaces [has,] in contradiction of all our theoretical notions, no advantage whatever." Their work also aimed to overturn what they called "Ritchie's Law of Surfaces." The law argued that activity increased "exactly in the ratio in which the surfaces of the two metals are increased." Tables reporting their results indicated the contrary.[32]

Their article, which displayed significant differences in battery activity depending on ratios of metals and length of time exposed to acid solutions, raised "the theoretical question of the source of electricity in the battery." The differences, they believed, extended beyond mere chemical inquires and into the realm of physics. William and Henry called on such researchers as Michael Faraday to examine further the reasons for the differences between theory and practice.[33]

Expanding on the useful arts theme of his article publications, Rogers published two books for use in higher education: one on engineering and the other on mechanical philosophy.[34] The engineering text offered an introduction to the subject for his students at the University of Virginia. It emerged from his lectures on the discipline, compiled in 1838 as *An Elementary Treatise on the Strength of Materials.* As with most of his other scientific publications, he made it a point to stress the relationships between the practical and the theoretical. To this end he organized the treatise around the physics of forces, theorems of comparative strengths and stiffnesses of materials, and sample applications of the theorems in practice. Calling it a "practical science," Rogers introduced his engineering students to basic concepts in weights and supports, compression and extension, forces and particles, all as they related to common "metals, woods, and other solid substances." He followed with a discussion of the physics of the forces of attraction and repulsion and the theories of particle movements within solid substances as a result. Rogers tied the abstract to the practical realities of the engineer and architect through the concept of elasticity, the capacity of solids to compress, extend, and return to its original dimensions. Understanding the capacities and laws of elasticity for a variety of materials, he noted, stood at the center of the engineer's responsibilities; the use of theorems eased these responsibilities.[35]

The *Treatise* also showed Rogers's use of an extensive base of research to support the principles of mechanical philosophy. He employed recent studies on the impact of weight on the compression and extension of matter, as well as "an accurate knowledge of the limits of elasticity" to inform his practical applications. Tables, for instance, provided students with lists of what was and was not known about materials to date. Rogers hoped that one day researchers would fill in the rest of the basic knowledge on materials of which "no accurate data have yet been determined." For over a decade after the publication of his *Treatise,* Rogers continued to revise and expand the work as new studies emerged in the literature. His familiarity with elasticity would reappear several decades later in a completely different realm: debates over evolution with Louis Agassiz.[36]

The second book, *Elements of Mechanical Philosophy,* differed from the first in providing Virginia students with an introduction to physics. In *Mechanical Philosophy* Rogers explained his view of the roles of various branches of scientific inquiry. Study of the natural world, he posited, took two predominant forms: natural history and natural philosophy. Natural history encompassed the fields of zoology, botany, mineralogy, and anatomy. Natural philosophy, meanwhile, included mechanical philosophy, chemistry, geology, and physiology. The former, Rogers suggested, was subordinate to the latter: "The leading aim of Natural History is the *classification of objects,* and it includes the study of actions and changes only as subordinate to this end. The great purpose of Natural Philosophy is the discovery of the *laws according to which the various changes of the material world are produced,* and it classifies actions only as a step in this result." The primary goal of *Mechanical Philosophy* was to investigate the laws, actions, and changes of matter and, for purposes of the useful arts, apply such investigations to practical ends.[37]

Thus, two main ideas appeared throughout *Mechanical Philosophy* that would inform Rogers's approach to science for the remainder of his scientific career. First, he made explicit his preference for a natural philosophical approach to science. Classi-fication, though once considered the dominant mode of scientific research, no longer represented the highest goal of the scientist, according to Rogers. "Observa-tion," he declared, "[is] not itself sufficient to contribute to science." The newer, more meaningful task of science, he asserted, involved the discovery and vigorous testing of laws of nature. Second, he divided the branches of science not by topic of inquiry but by methodology. Mineralogists, for example, worked as natural histo-rians with the method of collecting, identifying, and classifying minerals within an established hierarchy or nomenclature. At the same time, geologists worked as nat-ural philosophers, so Rogers argued, by investigating the laws that govern the in-terrelationships between minerals and such dynamic forces as heat or compression. Mineralogists and geologists, therefore, worked toward different aims through different means, however similar the material under investigation.[38]

Rogers elaborated on his sharp distinction between the roles of natural history and natural philosophy in an address at Williams College in 1855. In August of that year Williams College celebrated its twentieth anniversary and the opening of Jack-son Hall, a building dedicated to its Lyceum of Natural History. Most of the hall was constructed to house and preserve specimens collected by natural historians. Not surprisingly, the whole tenor of the occasion implied a celebration of collection and classification, but Rogers chose to focus his speech on the direction of science and natural history's relationship to changes then occurring. The address delivered at the celebration punctuated the differences he viewed between natural history and nat-ural philosophy, discussing the rise of new modes of inquiry that had rendered old ones obsolete or inaccurate. In particular, he gave examples of how classification by the external structures of flora and fauna, as performed by natural historians, could be shown to be inaccurate through the use of advanced microscopes and chemical analyses. The organic and inorganic distinctions, from Rogers's perspective of the emerging changes in natural philosophy, no longer appeared accurate. Such dis-tinctions, he remarked, were an obstacle to a "deeper scrutiny into Nature, to culti-vate a keener observation, and to consider the objects . . . not as they have been pre-viously described, but in their more vital and profounder relations." The physical sciences must pass beyond "the purely statistical view of living creatures." By "sta-tistical view" he was referring to the common practice of counting similarities and differences between organism as evidenced by mostly external structures and viewed with the naked eye. Chemical tests, he explained, had shown that some living forms are composed of both organic and inorganic matter and that classification systems needed to be revised accordingly.[39]

Rogers concluded that advances in chemistry and optics had changed the way scientists considered classification, for external structures now gave way to an understanding of internal structures and features unseen by the naked eye. "Without the refined appliances," he stated, "the almost infinite variety of forms and structures . . . would have been unknown or vaguely recognized." Characteristic of his ideal of the useful arts, Rogers highlighted the relationship between developments in technology and developments in science. As technologies allowed scientists more ways to observe the natural world, new questions and problems emerged. Should classification be based on the external structures of organisms and materials? Or should it be based on the object's reaction to chemical and mechanical changes? Or should the arrangement of particles as found under the microscope also affect classification? Rogers had faced similar questions across his own career, as innovations changed research in geology, chemistry, and electrical science. Considering the recent developments, he implied, natural history—its buildings, cabinets, methods, and ideals—had a noble past but an uncertain future. If his public ideas about science and education are any indication, he may have wished Williams had opened a laboratory that day rather than a museum.[40]

Not all scientists agreed with the distinctions Rogers made in his physics text and in the speech at Williams College. With few exceptions the scientific career of geologist and zoologist Louis Agassiz epitomized the classification-based science Rogers took issue with in his speech. Swiss-born Agassiz had gained a prominent reputation in Europe as a natural historian through publications largely on glaciers and in the area of ichthyology. After arriving in the United States and accepting a position at Harvard, he began his multivolume series *Contributions to the Natural History of the United States* (1857–60) with the kind of approaches to science that Rogers considered destined for replacement.[41]

These differences between approaches to science extended well beyond the mere organization of science and into the perception of advances in various fields. If scientists like Agassiz held fast to older models of scientific inquiry based on classification, they would perceive nothing but challenge and displacement from newer methods that overturned the kind of science upon which they had built their careers. From the span of Rogers's scientific career, no clearer example of the different approaches to science and their impact on the perception of advances in the field can be found than in a series of debates over evolution that took place in 1859 and 1860.

Rogers's main foil in the debates over evolution was none other than Louis Agassiz, but their debates in 1859 and 1860 were hardly the first discussions of the topic in America. Theories on evolution received attention from scientists well before the publication of Charles Darwin's *On the Origin of Species* in November 1859. Some scientists followed debates in France over the works of Jean Baptiste Lamarck or in England over the theories of Erasmus Darwin, grandfather of Charles, or Robert Chambers. Homegrown debates had even cropped up between Harvard botanist Asa Gray and Agassiz shortly before the appearance of *Origin*. The outcome of their intellectual jousting sheds light on Rogers's approach to the same topic.[42]

Asa Gray, a friend of Charles Darwin, understood some of the differences between *Origin* and preceding evolutionary writings, largely through correspondence with his English counterparts before the 1859 publication. Joseph Hooker at Kew in England worked on taxonomic problems similar to those taken on by Gray at Harvard. Gray's appointment in Cambridge had two parts: instruction in botany and "superintendence," or directorship, of the Botanic Gardens. Harvard placed an emphasis on the gardens work, allowing Gray ample time for correspondence with European scholars. Through Hooker, Darwin, and others in a small circle of confidants, Gray came to learn of the central tenets of *Origin:* natural selection, the struggle for life, the formation of species. He began to see certain patterns in his own work that matched or could be explained through Darwin's theories. In particular, Gray's comparative work on Japanese and American botany raised questions about the distribution of species. For Darwin the distribution had to do with natural selection, rather than divine placement. With this conclusion Gray challenged Agassiz in debates that provided an early introduction to ideas in the coming publication.[43]

Having been largely influenced by his training at the University of Munich and studies under Georges Cuvier, Agassiz embraced idealism of *Naturphilosophie* and the immutability of species. The idealism he absorbed at Munich through the teachings of Lorenz Oken and Friedrich Schelling provided him with a speculative framework for the common unity of organisms, the "great chain of being," to which he would later refer during his debates on Darwin. In his view on organismic unity, species originated from an idea of the creator, placed along a continuum from lowest to highest in order, capped by humans as the most superior of creative acts. In Paris, under Cuvier, Agassiz's views on the fixity of species crystallized. Indeed, he had attended the well-known 1830 debates between Cuvier and Etienne Geoffroy Saint-Hilaire, debates in which Cuvier crippled Lamarckian thought, closed the doors on evolutionary speculation in France for fifty years, and left a lasting imprint on Agassiz's thought. Convinced that evolution had failed in France, Agassiz saw no reason to entertain the ideas in America and promoted, instead, the intellectual heritage of idealism and the fixity of species.[44]

While much has been written about the Gray-Agassiz debates, few have considered the outcome and its relationship to Rogers. Put simply, the two men sparred over Gray's contention that the similarities between Japanese and American botany called into question Agassiz's view of locally specific acts of divine creation. But Agassiz brushed aside Gray's challenge for two reasons: one scientific, the other rhetorical. First, while Gray excelled as a botanist, his command of geological research was not particularly strong. To Gray's misfortune his challenge against Agassiz partly relied on a geological explanation for the migration of plants from America to Japan (or vice versa). Agassiz, more familiar with geological theory through research on glaciers, revealed obvious flaws in Gray's argument. Observers recognized Gray's geological shortcomings in the debates held in December 1858 and the spring of 1859 at Gray's Cambridge Science Club and at the American Academy of Arts and Sciences. Second, Gray lacked preparation in the rhetorical skills necessary to overcome Agassiz's "affability" and "charms." Agassiz had a facility in the art of persuasion that Gray lacked. Deficiencies in Gray's presentation abilities were recorded by one of his botany students. Although the student enjoyed the *content* of the lectures, "the manner was positively shocking. I never saw a person more awkward in delivery." Some scientists and science enthusiasts sympathized with Gray's position, but the debates did nothing to displace Agassiz prominence as America's interpreter of science.[45]

Rogers had strengths in areas in which Gray was weakest. A familiarity with European and American geological thought allowed him some flexibility in contending with the discourse surrounding evolution. Moreover, Rogers was neither an undistinguished speaker nor careless at building arguments. Students as well as colleagues and politicians noted his strengths in these areas throughout his career. Thus, Rogers had an opportunity to influence the reception of Darwin in ways that Gray could not. When Agassiz, who wielded significant influence from his professorship at Harvard, came to disregard Darwin's theory of evolution as a passing fad, few were willing or in a position to challenge him. Initially, he left Darwin's work alone, claiming it unworthy of critical attention. Yet, when the ideas contained in *Origin* did not fade from scientific discourse, Agassiz found himself working out an impromptu response during his encounters with Rogers.[46]

Ironically, Agassiz had once believed he had made a convert out of Rogers. When Agassiz first arrived from Europe, he claimed that Rogers and he maintained similar approaches to natural history. Agassiz viewed Rogers as standing prominently among "a most respectable contingent . . . [who] have long been familiar with European science." He assumed that he "had the pleasure of converting already the most distinguished American geologists to my way of thinking; among others, Professor Rogers." On such specific matters as glacial theory, they indeed shared similar ideas, but on larger questions of methodology, classification, instrumentation, the useful arts, and the role of natural philosophy, the two stood an ocean apart. By 1859 their commonalities paled in comparison to their differences. Rogers and Agassiz, by then, took opposite sides of the evolution debate. The deep-seated division became exposed in six unplanned debates leading to and centering on the publication of *Origin*. According to observers, the outcome of these debates—held at meetings of the Boston Society of Natural History and the American Academy of Arts and Sciences—depended not only on their initial interpretation of the work by way of their scientific background but also on their rhetorical skills.[47]

Agassiz first learned of *Origin* in November 1859. Darwin had sent him a copy and included an appeal for a fair hearing, even though he understood there were differences between them. "The conclusions at which I have arrived on several points differ so widely from yours," Darwin explained, "[yet] I hope you will at least give me credit, however erroneous you may think my conclusions, for having carefully endeavored to arrive at the truth." Agassiz's reaction was simple and explicit: "This is truly monstrous!" he wrote in the margins of the text. Whatever his initial reaction, Agassiz found himself in an intellectual tug-of-war over the matter with Rogers. Although Rogers first learned of *Origin* through correspondence with his brother Henry, who was then in Europe, he understood the basic arguments of the work and began to consider their merits. Unwilling to either dismiss or embrace Darwinian theory, Rogers left open the possibility of accepting the notions upon a careful reading of the work itself.[48]

The same month in which Darwin published *Origin*, Agassiz received a copy, and Rogers began learning of the fundamental precepts of the work, Rogers and Agassiz began their evolution-related debates. They focused on stratigraphy at first. Rogers presented work on fossil findings that had implications for redefining geological epochs and reconsidering the existence of the fossil species across periods layered in the earth's crust. He concluded that stratigraphic divisions could be more accurately understood if grouped in larger formations rather than more minute divisions. Agassiz perceived correctly a potential challenge to special creationism. The minute formational categories of geological epochs were central to Agassiz's explanation for the diversity of species. With each epoch came a divine intervention that explained how the organisms appeared wholly new. Greater numbers of epochs meant greater frequency of interaction between "creator" and creation and, thus, a potentially more plausible explanation for the multiplicity of species. Revising the epochs, creating fewer and larger categories as Rogers had proposed, conflicted with the idea of frequent intervention and struck at the core of Agassiz's fundamental assumptions about the origins of species.[49]

On one occasion Agassiz rose abruptly and questioned the findings, asking for "proof" of how one particular "series so-called is a single formation." He drew attention to the way geologists had at times divided eight or nine formations out of a single formation, "each with its characteristic fossils." Rogers replied by defending the term *series* to mean a collection of formations accumulated during a single geological period and countering with additional examples that challenged Agassiz's position on more minute distinctions. As for proof, Rogers had already made extensive use of and comparisons between evidence he had found in Nova Scotia and the Virginia survey.[50]

They also debated Darwin's work itself. Agassiz viewed *Origin* as little more than "fanciful theory." Evolutionary theory, he claimed, could not itself survive when considering a "fatal objection," a counterexample that Agassiz had brought forward. For Agassiz, if an organism, such as the *Lingula prima,* had survived essentially unchanged through several geological periods, then Darwin's hypothesis of "gradual development" was erroneous. Over time, Agassiz argued, the species should have changed or modified as described by Darwin's notion of natural selection. Yet, because it had not, the theory proved incapable of explaining the counterevidence. Rogers, drawing on his experiences as a geologist and natural philosopher, understood the objection well. For Rogers explained that geological evidence also supported the notion of migration in which more developed species may have changed settings and locales. Rogers also appealed to the audience, which some accounts describe as enthralled by the exchange, with an argument based on natural philosophy. He suggested that some organisms have "great energy of resistance [to change], and some very little." The fact that one species had persisted over time, Rogers asserted, did not mean that others had not changed during the same period.[51]

In addition to stratigraphy and Darwin's work, the two men squared off on specific terminology. Agassiz challenged Rogers's notion of species migrations by way of a discussion about the lowest system of fossils. In doing so, Agassiz made what some observers described as a confused statement on how the most complex organisms could be found in the earliest epochs and that the complexity and diversity seen in current fauna would be discovered in the fossil evidence of previous periods. The confusion centered on Agassiz's multiple uses of the words *low, high,* and *perfect.* Primitive fish embryos, for example, showed the highest and most complex nature, Agassiz believed, for they contained the essential matter for creating the most complex fish.[52]

In Rogers's opinion Agassiz had stretched his terms into meaningless categories, and he called on the Harvard professor to clarify. The problem with using such terms as *perfection* was that it proved "just as indefinite as the word 'species.'" If the "ancient types of life" showed marks of highest order, then it contradicted Agassiz's own assumption that humans represented the creator's most "perfect" creation. To this Agassiz closed the debate with a claim that "the vertebrate egg is superior to man himself, in as much as it embodies all that may be produced from it." Rogers recognized that Agassiz could not have it both ways and capitalized on the inconsistent use of terms and argumentation. Aimed at persuading many in the audience, Rogers provided counterexamples via his extensive knowledge of geological science.[53]

Their six encounters left Agassiz on the defensive. He had not expected the "immediate reception" of his ideas, but he remained "convinced that they were true." Despite the geological evidence and counterexamples from Rogers, Agassiz continued to hold that special creation would stand the test of time whereas Darwin's ideas would soon pass.[54]

Both debaters displayed a command of the scientific literature and of evidence. Just as important, however, were their rhetorical strategies. Agassiz assumed the role of Socratic questioner to "teach" his opponent and listeners a traditional view of natural history. He asked for "proof" from his opponent when considering new geological ideas. Rogers replied with lectures that outlined evidence from multiple geological surveys. In the midst of the debates Agassiz confused terms (i.e., *low, high,* and *perfect*) or contradicted himself. Rogers was quick to point out the errors. Agassiz offered "fatal objections" and analogies; Rogers turned such liabilities into advantages and examples of gradual development. Agassiz appeared dogmatic when resorting to ad hominem attacks (i.e., "fanciful") against Darwin's work. Rogers called attention to Darwin's own words and approach to science that suggested "fairness" and "absence of dogmatism."[55]

Observers close to both Rogers and Agassiz commended the debaters' knowledge of natural history and natural philosophy but emphasized this blend of rhetorical skill and logic as a determining factor in the outcome of the debate. Nathaniel Southgate Shaler, an Agassiz disciple, for example, gave the victory to Rogers. "Agassiz was admirable in discourse," explained Shaler, "but his capacity for debate was small. Rogers, on the other hand, was not only an able and learned geologist, but very skillful in argument, with a keen sense of the logic which should control statements." Shaler was hardly alone in his assessment. During the debates Rogers received correspondence from observers who made specific reference to his presentation style. One correspondent attending the sessions appreciated Rogers's handling of Agassiz's reversals, claiming to have come away from the debate "interested," "instructed," and "highly amused" by the performance. Jules Marcou, geologist and later editor of Agassiz's letters, came to a similar conclusion. Marcou commented that Agassiz "quickly lost patience, became excited, and showed signs of vexation." In the end "he was defeated" in the debates partly because of Rogers's rhetorical style. The Harvard professor had clout, however, and Rogers suffered some political consequences from the scholarly scuffle. "It would have been politic on his [Agassiz's] part," continued Marcou, "if he had offered the [Harvard] chair of geology to William B. Rogers, then a resident of Boston. But Agassiz did not like to have any one so near who might overshadow him." Rogers never received an appointment at Harvard, but his scientific thought—as characterized by the useful arts and by the debates with Agassiz over evolution—would over time continue to fuel a creative energy he possessed for imagining other possibilities, such as MIT, in the area of higher learning.[56]

As Agassiz the scientist began to fade from the scientific community, so, too, did his idealism and belief in the fixity of species. The Darwinian revolution appeared to reach even the closest aspects of Agassiz's life when his son Alexander and Edward S. Morse, another of many Agassiz disciples, became evolutionists. Morse recalled that his decision to abandon Agassiz for Darwin was informed by a "very interesting dissertation from Prof. Rogers" at a meeting of the Boston Society of Natural History. Rogers, of course, hardly toiled alone in the initial reception of Darwinism in the United States. Asa Gray, Jeffrey Wyman, and others, for instance, published reviews of the theory for lay, scientific, and religious audiences. But it was Rogers who exposed Agassiz's inconsistencies more clearly and powerfully to observers of the debates than Gray had accomplished to date. The proximity of the debates to the publication left no lag time for Agassiz to control the initial reception of *Origin*.[57]

The confidence with which Rogers engaged the debates on evolution owed much to both Baconian and Humboldtian traditions. Without the collection and classification work he had conducted on the survey of Virginia, Rogers would not have had the detailed knowledge of geology that became central to the debates. The basic Baconian-style science aided him in the largely theoretical dispute over the origins of species. At the same time, Rogers had also published on mountain formation theories and had developed research interests in natural philosophy reminiscent of Humboldt's "terrestrial physics." Particularly relevant to the debates were notions about the dynamics of elasticity, which Rogers used metaphorically to relate to the idea of changing, evolving, elastic species.

Clearly, the useful arts ideal was a central part of Rogers's scientific career. He directed geological research and physics experiments with a view to advancing practical as well as theoretical knowledge in the United States. Rogers made a conscious effort to reconsider theories or laws that failed to connect with the field or the laboratory; at the same time, he placed a significant emphasis on reaching the greatest number of people with data and concepts derived from science for practical ends. Rogers's desire to keep the channels of science open to the greater populace informed his ideas about the organization of the science profession itself.

CHAPTER FOUR

Advancing and Diffusing

A S A GEOLOGIST, more so than as a natural philosopher, Rogers participated in some of the earliest efforts to professionalize science in the United States. To those efforts he brought along the same useful arts ideals that colored his scientific research. From the survey of Virginia he learned the art of combining state interests in useful knowledge with the geologist's desire to gather scientific data. He believed that any initiatives for professionalizing American science must include the interests of the practical scientist as well as the theoretician. Geologists understood this better than most scientists before the Civil War, for it was their group that received uncommonly generous support for research. This support raised utilitarian and abstract questions from the start. State governments across the Northeast and the South opened their purses to support geological surveys and topographical mapmaking. Federal support launched explorations and surveys to the west fueled by the expansionist dreams of government officials, dreams rekindled with the acquisition of each new territory. Likewise, the most prominent scientific institution of the period, the United States Coast Survey, received unparalleled financial and political support. The coast survey lasted decades and would come to represent America's largess and favoritism toward the discipline.[1]

Rogers reflected on the spate of geological activity from the perspective of the useful arts, reflections that significantly informed his professional ideals. American democratic values had chiseled into the New World landscape a call for useful information that aspiring professionals of any sort could not ignore. Yet extreme interests on both ends of the practical-theoretical spectrum of scientists threatened the collapse of early professionalization efforts. Rogers and a circle of like-minded scientists worked toward resolving tensions that came from competing interests. They worked to keep science organizations and communities together. In this way Rogers not only witnessed the development of the American science profession but also participated in its formative, emergent, and crisis periods. Across these phases he advocated the advancement and diffusion of knowledge based on the ideal of the useful arts.

Rogers and other geologists led the professionalization of American science, in large measure, because of the economic support they had received. Government funded geological surveys of the 1820s and 1830s accelerated two vital conditions for professionalization: specialization and a demand for standards. The U.S. Coast Survey, for example, became well-known for giving advanced training to its researchers. Parents of young, aspiring geologists pleaded with Alexander Dallas Bache, during his tenure as director of the project, to accept their sons as assistants. The value of surveys for specialization and science training added prestige to the field; prestige, in turn, facilitated the geologist's claim to specialization. At the same time, the expanding number of surveys, directors, researchers, and assistants in geology brought to the surface unresolved dilemmas in the field. Old debates over nomenclature and other elements of geological research became pressing problems and obstacles to progress. With prestige acquired from large-scale projects and calls for standards across their work, geologists began to seek unity through professionalization.[2]

In 1840 representatives and directors from ten state surveys huddled at the Franklin Institute in Philadelphia to convene a new organization. Rogers kept abreast of developments that year through one of his survey assistants and two of his brothers, Henry and Robert. Professionalization, in terms of institution building, stood as the primary goal of the three-day meeting, marking the beginning of the Association of American Geologists. At the time few would have predicted that the little-publicized gathering would give rise to an icon of American science. The tepid support for or outright failures of previous societies and associations made most observers skeptical of any new efforts. Late-eighteenth-century organizations such as the American Philosophical Society (APS) and the American Academy of Arts and Sciences (AAAS) provided the geologists with potential models for organizing. But the models did not go far enough toward advancing and diffusing knowledge in the manner called for by specialists in geology. Rather, the APS and AAAS promoted a generalist approach to science that failed to address specific geological concerns. Early-nineteenth-century organizations such as the Columbian Institute for the Promotion of Science and the American Geological Society had problems as well. The former lacked the support of leading scientists, and the latter experienced organizational difficulties that caused its demise within a decade of its founding.[3]

Aware of the inadequacies of previous institutions, the Rogers brothers took note of the discussions at the 1840 meeting. Edward Hitchcock, president of Amherst College and director of the Massachusetts survey, proposed the formation of an American equivalent to the British Association for the Advancement of Science (BAAS). But he also recognized the difficulty of assembling scientists from diverse branches of science, and he knew of widespread doubts over the possibility of maintaining such an organization. In light of these reservations, Hitchcock argued for a smaller, geologist-based association. The idea met with approval from most listeners, including Henry and Robert. Henry's support, in particular, was critical to Hitchcock's efforts, especially given Henry's exposure to British associations during his excursion to Europe in the 1830s. When Hitchcock turned to Henry for advice, Henry consulted with William "on whether it were better to delay the movement until a General Association for all the sciences can be brought about or to make it now for geology merely." Consistent with his later policy positions, William most likely recommended slow growth with the goal of ultimately achieving both aims. In 1842 he took such views with him to an organizational meeting that adopted a constitution resembling one established by the BAAS. The members decided to include more than geologists, and the name was changed that year to the Association of American Geologists and Naturalists (AAGN). They stated their central mission as "the advancement of Geology and the collateral branches of Natural Science." This effort for advancement included "the promotion of intercourse between cultivators of science" through annual, migratory meetings.[4]

Rogers took part in the development of the AAGN in its early years and enjoyed professional and personal satisfaction from the experience. He gained professional satisfaction from the opportunity to discuss at the first several meetings specific geological concerns. These discussions focused on developments in American and European research, on bringing consensus to much disputed geological terminology, and on developing an outlet for publication. Rogers expressed satisfaction with these developments irrespective of his own success or failure at introducing new ideas into the American science community.

On some occasions Rogers received recognition from the association for his work on the Virginia geological survey. Hitchcock's presidential address to the organization in 1841 lauded Rogers's exemplary work in "microscopic paleontology." Hitchcock mentioned the "interesting discovery by Prof. W. B. Rogers" involving minute chemical analyses of the tertiary strata of Virginia. "If such is the beginning," he asked listeners, "what, gentlemen, will be the end of this infinitesimal geology! We seem fast advancing." Similar praise went to Rogers the following year for a presentation on approximately thirty thermal springs in Virginia. Data on the springs from the state survey included a wide range of temperatures and chemical compositions collected at various sites. Location, Rogers argued, especially with regard to placement along the Appalachian chain, had great bearing on the contents of the springs. "These, it is believed," mentioned one listener, "are the first developments of the kind made in the United States, and, if we except those of [Alexander von] Humboldt in Mexico, the first in North America." Certainly, Rogers took pride in sharing his work with aspiring professionals in his field.[5]

On other occasions commentary on Rogers's work was more critical and sometimes combative. His notions about geological nomenclature that emerged from collaborations with his brother Henry, for example, received a cool response from his colleagues. In a presentation on "a system of classification and nomenclature," William and Henry advocated a numerical pattern for identifying the earth's layers, a pattern based on studies of the Appalachian mountain range. When the brothers finished their presentation, geologist James Hall, promoter of a competing nomenclature based on local names from his survey of New York, applauded the general contributions to geology while discounting the Rogers number system. William shot back that Hall had missed the point, that he "had not . . . understood . . . the intricate structural geology" and their proposal for an alternative to local names. As one observer of the debate recorded, "Mr. Hall replied that the term 'New York system' of rocks was considered by the gentlemen who agreed on it, as a convenient conventional term." Although the Rogers brothers "objected strongly to any nomenclature, based upon an examination of local districts . . . Hall and others advocated the more cautious method" of the local New York system. Edward Hitchcock also challenged Rogers's work for not using the New York system or a recognized system from Europe. While acknowledging the "vast series" covered in the Appalachian studies, Hitchcock questioned the use of numbers to explain the mountain formation without regard for other known systems. "Whatever may be their views as to the identity of these groups," he stated, "with rocks described in other parts of the world, they have refrained from expressing an opinion, in their annual reports." Rogers believed it was too soon for U.S. geologists to compare their formations with those of Europe. Hitchcock disagreed.[6]

The mixture of compliments and critical commentary did not sour Rogers on the idea of professionalism. To the contrary, he viewed the AAGN not as an organ for the promotion of his own ideas but, rather, as a forum in which to settle scientific controversies. Of these meetings Rogers remarked on his "feeling of great satisfaction . . . at the straightforward devotion to science which had marked so strongly all the proceedings of its members." Notable visiting members, such as British geologist Charles Lyell, offered commentaries on the meeting's presentations and debates, giving the fledgling organization added confidence. Moreover, attendance by the general public infused the sessions with popular support. For one address by Yale scientist Benjamin Silliman, an estimated five hundred observers came to listen.[7] Each year Rogers, who looked forward to distinguished visitors and well-attended addresses, derived professional and personal satisfaction from being with friends of science: "For us such reunions of the scientific brethren as our Association of Geologists are of precious value and form the best compensation we can enjoy for the prolonged restraint of our vocation. What new impulses of exertion, what encouragement and guidance do they not give? And then in our hours of lonely meditation to how many cheering and delightful social recollections."[8] The professional and social value of the AAGN strengthened the resolve of members to keep the organization stable.[9]

Rogers aided in bringing stability to the organization by continuing to present papers and by holding offices. Presentations were a valued resource to its members. As with debates such as those over nomenclature, the presentations at the AAGN meetings allowed for discussions on topics under dispute. The association also published its proceedings, making public the state of geology in America. A sense of nationalism prompted members to take the presentations seriously, for it was their work that might be read in Europe. The Rogers wave theory of mountain chain formation presented before the AAGN, for example, received attention in European circles. Although the theory itself had a short shelf life, William's brother Henry Rogers traveled to Europe in the 1840s and found that fellow geologists had taken an interest in "the labours of William and myself in geology, and the fraternal association of our names." Some scientists had "read all we have written, and even said, at the meetings of our Association." Advancing knowledge through research and diffusing knowledge through presentations had much to do with membership invitations received by William from the Geological Society of London and the Royal Society of Northern Antiquaries of Copenhagen.[10]

The scope of Rogers's presentations mirrored his interests in science, spanning topics in geology and chemistry. Many of the papers had to do with his survey work and the useful arts approach he brought to it, which included practical and basic research interests. The practical papers often related to technologies he had devised or modified. Better technology in the United States, he argued, allowed for greater precision in chemical findings. "By a peculiar form of apparatus [designed by Rogers], furnishing very accurate results," remarked one observer, a "saline solution has been found to absorb a far larger proportion of carbonic acid, than is attributed to it by Professor [Justus] Liebig." Other research technologies came directly from Rogers's survey work. With his brother Robert, William constructed a new "apparatus" to determine the amount of iron in iron ores and cast iron. Both laboratory innovations aimed at improving precision in chemical and geological research.[11]

Basic research presented by Rogers often described rare geological findings in Virginia. He reported to his colleagues on the extreme thickness of coal formations in the eastern part of the state. The unusually thick strata near Richmond, Rogers told listeners, reached "upwards of eight hundred feet" in some areas. He compared the strata with formations known in the west and in Europe and "laid much stress," as one commentator described, "on this determination as supplying one of the links in the geological series not hitherto discovered in this country." The work on coal strata reflected developments in the uses of fossils to identify strata relative to the layer's age and placement among other layers.[12]

In addition to preparing papers, Rogers held offices in the organization. He served twice as president of the AAGN. During the first tenure in 1842, in which he replaced Samuel G. Morton as chairman, Rogers was buoyed by the meeting's strength in numbers and directed the association to continue a plan of steady growth. He later commented to Benjamin Silliman, a past president, that he looked forward to having the AAGN "enlarge its ranks while expanding its usefulness and reputation." When Rogers assumed the presidency for a second one-year term in 1847, his hopes for the organization had become realized. That year's meeting held in Boston generated large crowds. Agassiz, often a crowd pleaser, felt compelled not only to offer commentary on the work of others but also to present his own papers at the conference. No longer an organization exclusively for geologists, the AAGN had developed into what the founders had originally desired for the American scientific community.[13]

As Rogers stepped down as president during the 1848 meeting, he guided the transformation of the AAGN into the American Association for the Advancement of Science. For most of the first day of meetings for the new organization, Rogers "presided at the organization" and inaugurated the "Constitution and Rules of Order," as drafted by his brother Henry. William offered a resolution that organized the association into two sections, one of them covering "General Physics, Mathematics, Chemistry, Civil Engineering, and Applied Sciences generally," and the other including "Natural History, Geology, Physiology, and Medicine." Thus, he served as chairman for part of the first AAAS meeting until he introduced William C. Redfield, president-elect for that year.[14]

After handing over the reigns to Redfield and leaving for a honeymoon in Europe with Emma Savage, in 1849, Rogers's interests in professionalization continued unabated. He took careful note of the meetings he attended overseas. The British Association for the Advancement of Science held its annual meeting in Birmingham and invited him to attend during his overseas visit. In Birmingham, he reported, "Darwin, Ansted, Ramsay, Mallet, Oldham, Griffiths and, above all, Murchinson, Sedgwick and Phillips among the geologists [are] taking me cordially by the hand." At the meeting Rogers presented formal "remarks connected with Murchinson's paper on gold veins" and attended the chemical section of the BAAS, where he discussed his research on the Appalachian springs. Concluding his visit with the British association, Rogers delivered an hour-long discourse on the geology of Virginia. Geologists such as Murchinson, De la Beche, and Lyell, he recounted to his brother Henry, complimented him for his "contributions to geology" and mentioned the wave theory as "having thus furnished a clue to the most difficult problems in European geology." Professionalization had helped to publicize general developments in the United States and, in particular, the scholarship of the Rogers brothers.[15]

All the while in Europe he made a careful study of professionalization in the Old World for ideas to bring back to the New. The most important advantage that American science enjoyed, he noted, was the "democratic" quality of participation in meetings, the free exchange of ideas between scientists of diverse backgrounds. "Oh, how happy should we be in America," he cheered, "in that security and sanctity of personal rights and free progress which we enjoy!" Although the British benefited from a more developed system of professionalization, American scientists, he believed, had greater opportunities for advancement. Because the European "men of science are poorly paid" and have "as a class an inferior social position," Rogers felt pride in the political and social freedoms back home and saw in them a boon for progress.[16]

The experiences in Europe influenced Rogers's pattern of involvement with the American Association for the Advancement of Science. During a period between its founding in 1848 and an annual meeting in 1851, most members anxiously desired to see the organization secure a permanent role in American science. Previous ill-fated attempts to found science organizations made many of them cautious, and no member wanted to witness another still-birth in professionalization. The first task of drafting a unifying constitution fell to Henry Darwin Rogers, who consulted with William on several occasions. First, having had experience with the British science community, the brothers recognized the need to adopt policies suited for the American context. Meetings of the British Association for the Advancement of Science had revealed to them a hierarchical order of membership, with voting privileges exclusively reserved for officers. Moreover, to receive membership in the BAAS, scholars first had to gain acceptance in other reputable societies. By doing so, the British association hoped to bring together the English and form a tightly bound coterie. The Rogers brothers rejected these elements of the European model when they considered the AAAS constitution. At the center of the American version stood the goal of uniting the American science profession through an inclusive approach to membership. The open policy during the formative years of the AAAS yielded a significant number of applicants, and only in rare cases did the association reject an application. Thus, the AAAS constitution placed few official requirements on membership. Any field researcher or college professor of science could apply, and if an applicant without such credentials received the support of an established member, the requirement could be waived. Indeed, the new rules made explicit that even "Civil Engineers and Architects" could apply for full affiliation with the AAAS.[17]

Discussing a final draft with William, Henry prided himself in making the constitution "democratic, federal, flexible and expansive, progressive, with all the true conservatism these features imply." The new constitution ultimately provided flexibility that would be useful for the organization's expansion. Although some of the more prominent members, such as Benjamin Peirce and Louis Agassiz, had reservations about the "democratic" ideals—particularly if perceived to support any science, including the practices of charlatans—only gradually did dissent emerge. For most of the formative and emergent years, members witnessed a rapid growth in membership, rising levels of participation at the annual meetings, approval from the public, and general internal harmony. Most every member agreed with the basic tenets of the Rogers constitution, characterized by the dual goals of advancing and diffusing knowledge. But such harmony wouldn't last forever. William would later come to champion the useful arts ideals in a constitutional crisis within the American Association for the Advancement of Science.[18]

During the 1850s Rogers's greatest challenge in the professionalization movement arose during a crises over the AAAS's mission. While he advocated the dual mission of the advancement and diffusion of knowledge, other members began to desire a focus on one or the other. A circle of geologists, inherited from the preceding AAGN, wanted to continue a course that emphasized the advancement of American science through specialized research. Their original concerns over geological nomenclature had not yet been resolved, and their efforts to search for a consensus created an increasing tension. Rogers and other elected officers attempted to strike a balance between specialists and generalists, for attention to the interests of generalists led to greater public approval and increased membership. In short, the dilemma stood between advancement, which brought prestige, and diffusion, which attracted popular attention and funding.[19]

Rogers, along with other officers and committee members, managed the dilemma in two ways. First, the association continued its policy of migratory meetings, ensuring for itself a national character and constituency. Local preparations at the meeting sites encouraged speaking engagements for popular scientific lectures at regional societies and civic group meetings. Yet the AAAS retained a sizable number of prominent scientists, a circle of scholars who assumed positions of authority within the organization. Second, the leadership began to tighten the policies for presenting work at formal gatherings. A resolution passed in 1851 declared that "no paper be read before the future meetings of this Association unless an abstract of it has been previously presented to the Secretary." The resolution marked a change in the open policy the Rogers brothers had helped establish for the formative and emergent periods of the AAAS.[20]

As the association began to stabilize, its leadership asserted greater control over the course of professionalization. Rogers became concerned that some form of oligarchic control might develop. After his involvement in a controversy at the 1853 meeting in Cleveland, his concerns grew. Jehu Brainerd, an Ohio amateur scientist, delivered what some called a wildly speculative paper on pebble formation. Geologist and standing committee member James Hall rose after Brainerd had completed his talk and dismissed Brainerd's research, motioning for the paper's exclusion from the proceedings of the meeting. The events marked the AAAS's first public debate over the criteria of acceptable and unacceptable scholarship. Brainerd defended his work, and for the following sessions the policies governing whether to publish his paper became a focus of concern within the organization. On one side members sought the right to present at the association's meetings and publish in its proceedings without interference. On the other side the AAAS leadership believed that amateur science had no place representing the United States abroad. Spontaneous debates surrounding Brainerd's defenders and opponents failed to solve the problem. During subsequent meetings Rogers and a special committee of nine other members tasked to "revise the constitution" grappled with problems of acceptable research and membership. In the mid-1850s, as hostilities toward the governing offices continued to mount over the rights of members, Rogers would serve a crucial role in the controversy.[21]

At the center of the crisis stood Rogers and his challenge to a small group of scientists led by Alexander Dallas Bache. By the 1850s Bache had established a formidable reputation in American science through the U.S. Coast Survey. Through his work he'd sharpened his abilities as an organizer, as an advancer of scientific knowledge, and as a cultivator of federal patronage. For many aspiring scientists Bache provided a model for research and promotion in American science. Solidifying his standing within the science community, he delivered an address before the AAAS during his presidency of the organization in 1851 that discussed the need to raise professional standards and to distinguish scientific excellence from quackery and charlatanism. The association, he argued, could promote such standards by cultivating greater support for basic research and avoiding the popular demands for utilitarian science. The rallying cry in the address focused on the great obstacle of insufficient funding for research. In short, Bache believed that, more than anything else, American science needed greater funding for its advancement. He stated that what was required were "rewards for principles, instead of for application."[22]

Rogers, in many ways, had a different emphasis that was informed by his experience with assistants on the Virginia survey. Certainly, both scientists shared an interest in advancing basic research. Rogers had displayed this interest in his survey reports and theoretical work on mountain chain formation. Moreover, Rogers had no qualms about promoting American science through funding agencies. His own experience on the state survey revealed to him the need for maintaining a steady source of support during the course of a multiyear project. But Rogers had a basic disagreement with Bache in that Rogers believed funding of research alone was not the greatest need. Rogers's experience in the field had taught him that the lack of qualified assistants posed a more immediate challenge to the advancement of American science. Without a corps of competent assistants, much of the research funding from whatever source—whether state, federal, or private—would largely be ineffective in advancing scientific knowledge. On the Virginia survey one of Rogers's constant challenges came in training and retaining field researchers, especially those with modest backgrounds in geology. He thus began looking for solutions to these problems by way of higher education reform. Bache, of course, was no opponent of new models for colleges and universities. He'd placed his prestige and power behind the idea of a university in Albany that would promote research among its faculty. But the Albany plan had failed in part because Bache and his circle hadn't received steady support for the enterprise. While they ran a sprint and failed, Rogers began a marathon; over time, through the founding of the Massachusetts Institute of Technology, he showed greater determination and dogged persistence in supporting the expansion of science in American colleges and universities.[23]

Bache and Rogers also disagreed with each other over how the AAAS should be organized. Bache pressed for major changes to advance his cause. He first advocated the establishment of permanent officers for the association. To provide for a more cohesive structure, the permanent officers assumed certain administrative details to ensure consistency between annual meetings. Bache also involved himself personally in the formation of a standing committee. Rogers and other leading members of the AAAS would not have disagreed with the notion of a standing committee in principle. The idea itself seemed palatable to those like Rogers who saw the need for such a committee charged with the purpose of raising professional standards. The committee, once established, began to direct association policy and assumed control over subspecialty section meetings. When, however, the newly expanded powers began to interfere with the functions of section leaders who decided which papers to accept or reject, opposition began to form. The lead dissenter was Rogers.[24]

It was not easy for Rogers to challenge Bache's personal agenda for the AAAS, for Bache did not work alone. As Rogers well knew, Bache led a small circle of self-proclaimed elite scientists. Within the circle and among its friends, they described themselves as the Florentine Academy and, later, the Lazzaroni. The term *Lazzaroni*, meaning "beggar" in Italian, reflected the mission this group had placed on itself: to beg for money for American science. In several ways the group succeeded in advancing its goals: Bache directed the coast survey, Agassiz dominated Harvard's Lawrence Scientific School and the Museum of Comparative Zoology, and Joseph Henry headed the Smithsonian Institution. Other distinguished members included Harvard's mathematical astronomer Benjamin Peirce, chemist Wolcott Gibbs, and astronomer Benjamin A. Gould. Each of them had well-established reputations, and together they formed a powerful coterie within the movement to professionalize science.[25]

To opponents of this emerging circle, the Lazzaroni were known as the "Washington-Cambridge clique" and as the "mutual admiration society." Rogers, in particular, resisted their attempt to retain power almost exclusively within their group. The presidencies of the AAAS, for example, had circulated among the group's members and friends since the association's founding. It became clear that they had managed to consolidate and perpetuate their power over the organization. The Lazzaroni managed this through the standing committee. From year to year at least four of the six seats on the committee were filled by the circle and its supporters; the other two seats had little power to challenge policies that the controlling members proposed. Through this arrangement, whether it was explicitly intended or not, the Lazzaroni controlled the power to elect association officers and members of the standing committee. The circle had assumed these powers since 1850, then largely unopposed when disagreements and controversies gave way to a desire for unity. As the association began to stand on surer footing, however, dissent began to emerge.[26]

Having played a part in shaping the original constitution for the AAAS, Rogers viewed the Lazzaroni as illegitimate usurpers of power. According to him, the constitution had explicitly accounted for elections by "ballot." Rogers interpreted this clause to mean that officers would be elected by the full membership of the AAAS, rather than by a select and powerful standing committee. His interpretation attracted attention from others who perceived that their role and power as members had declined since the rise of the Lazzaroni. Geologists, in particular, recalled that they had founded the AAGN, the precursor of the AAAS, which at that time centered on the concerns of geological problems. The sudden rise to power of Bache's circle left a core of founding members disgruntled. Some went as far as to advocate a separate society. Although Rogers took an interest in the complaints of geologists, he chose reform rather than a splintering of the AAAS.[27]

Drawing from a base of support, Rogers began to challenge the Lazzaroni in the mid 1850s. When the Bache circle decided to appoint a special committee to revise the association's constitution, tensions escalated between the standing committee and the general membership. The Lazzaroni appointed a constitution committee to quell the dissent. Not surprisingly, however, most of the members assigned to the task came from the ranks of the Lazzaroni and its close associates. Recognizing Rogers's popularity with dissenters, the constitution committee selected Rogers as a member, a token gesture to the general membership. What the Lazzaroni might not have realized was that his commitment to the useful arts ideal would frustrate their attempts to expand their control. Through the committee Rogers expressed two major concerns: the use of the AAAS for the promotion of personal projects and the procedure for electing the association's officers. The first concern implied that the association had turned into a self-congratulatory platform for a select few. Rogers disapproved of the advertising of personal projects, calling it propaganda rather than research; he wanted to add a rule to the constitution for "precluding all action in the way of recommendation or otherwise, either of instruments, books, researches, or other scientific, public, and private enterprise." The Bache circle balked. They argued persuasively enough to have the rule discarded, defending the association's function of appealing to funding agencies. Rogers's first challenge fell apart.[28]

On the other issue, the election of association officers, Rogers enjoyed more success in challenging the Lazzaroni. Indeed, his efforts limited their control and ultimately altered policies that had given them nearly exclusive governance over the association. Their downfall began in 1855, when the committee to revise the constitution issued a majority report recommending changes to the original version. Great controversy arose over the obvious attempt by the Lazzaroni to solidify their control. One clause read that the expanded powers would include the right "to assign papers, arrange the business, suggest places and times of meeting, examine or exclude papers, appoint the local committee, nominate persons for membership, and decide on publication." Reporters from the *New York Times* and the *Providence Journal* observed that the report had dropped like a bomb on the meeting with the potential to break the association apart. Tension had risen to a critical stage with Rogers standing between the constitution committee and the general membership. He believed that the true intentions of the Lazzaroni had been exposed at the meeting, that the full membership better understood the nature of the ensuing power struggle; thus, Rogers contented himself with a minor victory when the committee decided to postpone a vote on the majority report. He then began to prepare for a protracted struggle expected at the following year's meeting.[29]

In 1856, when the standing committee voted to install more Lazzaroni members into AAAS offices, Rogers led a protest against the procedure. The debate centered on whether the officials were to be elected by the standing committee or by the general membership. He shared the details of the events with his brother Henry as they unfolded, stating that "the constitutional question was as usual staved off as late as possible by the Bache party, but I compelled them at last to bring out their propositions, as to the power of the Standing Com, in a distinct form, as the Majority Rep. of the Const. Com; and I confronted it by my Report, or that of the Minority." Once the two reports were placed before the general membership, "the Majority report was stricken out by a large vote, and mine was then substituted for it." "Thus, at last," he explained,

> with all their efforts, and in spite of the numerous dependent votes controlled by this party, they have been entirely routed in the Assoc. and we may count upon a fair representation of the general interests in the Standing Coms. The Majority Report proposed . . . the usurpation they have been doing of late years. My Report proposed that they be elected by ballot on open nomination from the Assoc. on their first meeting. . . . The Coast Survey and its alliances were at first quite sure of legislating their usurped powers, and their defeat in this matter has been a serious blow. As I had to take the lead in this opposition, and spoke very earnestly, though respectfully, I have come in of course for a principal share of their indignation.[30]

In this struggle over association policy Rogers championed the cause of the general membership. But he did not face the Lazzaroni completely alone. He claimed as "chief helpers" Ormsby M. Mitchel of Cincinnati, Chester Dewey of Rochester, Edward Hitchcock of Amherst, W.H.C. Bartlett of West Point, and "many other of the old and highly respectable members."[31]

The most flagrant opposition, Rogers told his brother, had come from geologist James Hall. As chairman of the organization that year, Hall had made his support for the Lazzaroni clear. In his presidential address he stated that "it is for the advancement of science, and not for its diffusion, that we meet. It is in the hope of communicating new truths, and adding something to the common stock of knowledge that we have associated ourselves." Rogers recognized the importance of advancement while also appreciating diffusion. His disagreement with Hall translated into the organization and procedures of the association. Thus, when Rogers won his victory over the Lazzaroni, Hall "acted the illiberal partisan throughout, and has been severely handled in the press for his incapacity and partiality in the chair." Encouraged by the victory, Rogers placed his hopes in the useful arts ideal "to maintain some control over the combination which has been so arrogantly claiming to regulate the Assoc. & with it the science of the country." Securing a place for both practical and theoretical interests, he believed, would only succeed if no single group controlled the organization's leadership. The Bache circle did not easily forget Rogers's challenge to its authority or his support for the useful arts ideal.[32]

Activities in the professionalization of science were hardly the only instances Rogers would cross swords with the Lazzaroni. Speaking to a gathering in New York in 1856, Bache asked, "Where is our American University?" As if responding to a similar query decades earlier, Rogers had begun proposing the idea of a technological institute.[33]

CHAPTER FIVE

Thwarted Reform

ROGERS'S IDEAS ABOUT higher learning followed the pattern of his scientific and professional thought. A combination of theory and practice stood at the forefront of what he believed a college or university should promote. The useful arts, in other words, appeared once again as an organizing principle in his worldview. During the antebellum period the United States underwent a scientific-industrial revolution that stimulated Rogers and others to translate their worldviews into college-level reforms. Rogers believed that one of the greatest challenges reformers of his era faced was the anemic role of science in academia. Charles W. Eliot, best known for his presidency at Harvard University from the mid-nineteenth to the early twentieth centuries, agreed with Rogers. In an article of 1869 Eliot evaluated the three most common ways science had entered the college curriculum: courses, separate schools, and technological institutes. At the start of the century colleges rarely offered more than a classical curriculum that required students to take Latin, Greek, and moral philosophy. As topics in chemistry or geology or astronomy generated interest, faculty began to provide occasional lectures or formal classes to students as electives. In addition to scattered offerings, the founding of separate schools provided another avenue for science in traditional higher education. Harvard and Yale established a popular model, continued at several other institutions, that separated classical from scientific studies. For students disenchanted with scattered courses or separate (or, as Eliot would say, "marginalized") schools, technological institutes provided a structured alternative. By the 1860s Rogers, Eliot, and other reformers believed that independent institutes held the most promise for scientific studies.[1]

In the decades before the Civil War, Rogers immersed himself in each of the three developments. All the while, however, he became increasingly known for advancing the idea of a comprehensive institute, one that aimed for scientific breadth and depth as well as laboratory instruction. His experiences in traditional higher education and scientific research had much to do with his vision for an alternative. Established colleges and universities, he came to believe, suffered from traditions that placed restrictions on science studies and favored a classical curriculum, which had long dominated American higher learning. After attempting to follow the path of reform, Rogers turned to institution building. His experiences conducting scientific research supported his contention that an independent institute was necessary. The lack of experienced assistants for the survey of Virginia illustrated to him the need for a new form of higher learning along with new modes of instruction. Few traditional institutions offered students regular or direct contact with the laboratory, and Rogers made this lack of laboratory practice the focus of his efforts for an institute. He based his useful arts worldview on the need for it.

In the early nineteenth century, if college students wanted science instruction, they looked not to the college but to medical schools. Rogers's father, among hundreds of others, graduated from the medical school at the University of Pennsylvania in the first decade of the century. Many of these students, in turn, either entered the medical profession or sought the rare college professorship in science. Benjamin Silliman followed this career pattern and became an influential science professor at Yale in 1802. For many of Rogers's generation, Silliman's chair represented a substantial fissure in the classical curriculum. It offered a model for the promotion of science in established classical colleges. Well-known scientists, such as geologist Amos Eaton, joined the swelling ranks of students who had studied with Silliman. Moreover, the highly regarded *American Journal of Science,* founded by the Yale professor, was known as "Silliman's journal." The journal provided a resource for scientists to keep informed of developments in the United States and abroad. Like other antebellum scientists, Rogers benefited from Silliman's efforts. By the time Rogers was beginning his career in science, Silliman had made professorships in such disciplines more attractive to college leaders. The prominence that Yale had achieved for hiring and keeping Silliman, America's science educator, drew other colleges into a race to stay current.[2]

The start of Rogers's career coincided with the first groundswell of experimentation in college science that occurred during the 1820s. At the time course catalogues began to appear regularly, and college presidents frequently compared their curricular offerings and admissions requirements. Advances made at one institution would follow shortly at others. For many reasons, including the common introduction of science into parlors, primers, and public schooling, college officials began to see the need to overcome obstacles to science instruction at the undergraduate level: the poor materials and textbooks, the lack of qualified faculty, the tenuous part-time positions for scientists, and the lack of apparatus for experiments and lecture demonstrations. Overcoming some of these obstacles, Harvard instituted its first regular chemistry courses for undergraduate students in 1825. A year later the faculty at

Amherst called attention to the "inadequacy of the prevailing systems of classical education" for ignoring science. Shortly thereafter, Amherst established courses on "Chemistry and other kindred branches of Physical sciences by showing their application to the more useful arts and trades." Throughout the 1820s additional calls for changes to the curriculum came from across the nation, from the University of Vermont, the University of Virginia, Columbia, Williams, Middlebury, Dartmouth, Dickinson, and others.[3]

At William and Mary, before decade's end, Rogers found little difficulty in making popular his lectures on natural philosophy and chemistry. By the time he replaced his father on the faculty, the college was rethinking its classical curriculum, particularly in light of the founding of the science-friendly University of Virginia in 1825. "The establishment of the University of Virginia," warned a faculty statement outlining the threat, "did not accord with the views of William and Mary, and it was foreseen that it would reduce its standing, unless some expedient was adopted, which might give a great impulse to the College." Science courses provided one "expedient" with which the institution could respond to change.[4]

Rogers brought to the lecture halls a generalist approach to science and presented the material in a manner that conformed to the useful arts ideal. In his natural philosophy courses he carried his young listeners through the topics of "Dynamics, Mechanics, Hydrodynamics, Pneumatics, Acoustics, Optics, Magnetism, Electricity, Meteorology, Physical Geography, [and] Physical and Descriptive Astronomy." But he also commanded student attention by illustrating the practical application of natural philosophy to "the strengths of materials, the construction of . . . Roofs, Arches, Bridges, Roads, the Steam Engine, and Elementary Principles of Architecture." Reflecting his useful arts approach, he relied on two basic texts. Students were required to read Rogers's *Introduction* to the field. His later publications, such as *An Elementary Treatise on the Strength of Materials* and *Elements of Mechanical Philosophy*, likely evolved from the text. Students also studied *Elements of Natural Philosophy* as selected from the *Library of Useful Knowledge*. Published in London, the *Library* provided a series of works that shared Rogers's useful arts worldview.[5]

In his chemistry courses Rogers applied a similar pattern. One section dealt with the theories of "Inorganic and Organic Chemistry," while another included demonstrations on the chemical relationships in "the arts of Bleaching, Dyeing, Tanning, Metallurgy, Brewing, Distillation, [and] the manufacture of glass and porcelain." The basic principles and practices that Rogers taught came from John White Webster's *Chemistry*, a work based on William Thomas Brande's text on the subject. Webster, an itinerant lecturer at Harvard, West Point, Brown, Amherst, and other colleges, had compiled his lectures and those of English chemist Brande for the volume, which "endeavoured not to limit the student to any particular theories, but to sketch the outlines of those of the most eminent writers, leaving to the teacher the discussion of their various merits." The review of theories came in large measure from Brande's own surveys of the field. A lecturer at England's Royal Institution, Brande received recognition for his theoretical research in the field, earning the prestigious Copley Medal in 1813. At the same time, Webster's coupling of theory with practice must have pleased Rogers: "Many valuable practical directions have been introduced from Mr. [Michael] Faraday's late work on *Chemical Manipulation,* and the section on the analysis of Minerals from Dr. [Edward] Turner's *Elements.*" To a large degree the text satisfied Rogers's useful arts emphasis for his classes on chemistry.[6]

His course offerings and selection of texts played a role in a heightened interest in science at William and Mary. One of Rogers's pupils, his youngest brother, Robert Empie Rogers, commented that "his classes are advancing very well indeed, and they are all very much pleased." Even for his evening study "clubs," sessions that William assigned to his classes and visited almost every night of the week, "students attend with the greatest alacrity possible; there is not the least disorder among them."[7]

The lack of disorder stemmed in part from Rogers's approach to instruction, which included laboratory demonstrations. This offering made his science offerings potentially more inviting, especially to students accustomed to the drone of recitations in ancient languages. He drew listeners in with his emphasis on technologies for instruction. Rogers took pride in using new approaches to college science "in a manner agreeable to the class." At the same time, he noted problems experienced with attaining and maintaining scientific apparatus. In his words, he found grave "difficulties arising from want of instruments, or from imperfection in those we possess, or any other trivial circumstances connected with my duties." Although the technical problems caused him "uneasiness or perplexity," he felt compelled as a scientist to "employ every accessible means of illustrating my subject in an intelligible manner." As a last resort, he could always rely on lectures, or "explanations," as he put it. Yet "the want of apparatus," Rogers concluded, "is certainly a serious difficulty in the way of a lecturer."[8]

Ironically, his efforts at the college led the University of Virginia to offer Rogers a position in 1835, an offer he felt he couldn't refuse. The faculty at the university supported an elective system that allowed Rogers more freedom to expand science offerings in the curriculum. From the moment he began teaching in Charlottesville, his practical and theoretical interests appeared in the institution's catalogue. At first the offerings were presented in scattered, unorganized themes. The first of four themes, Mechanics, included "Statics, Dynamics, Laws of Impulse and Pressure, Corpuscular Forces, Strength of Materials, Friction and Machinery." Many of the topics lent themselves to practical uses of physics for engineering purposes. The second section covered "Hydrodynamics" and the third "Heat." The remainder were combined into a theme with assorted topics, such as "Electricity and Galvanism; Magnetism; Electro-Magnetism; Optics; Astronomy." Within a few years, however, Rogers reworked the offerings into an organized system that followed the scheme adopted in his *Elements of Mechanical Philosophy.* Until he published the work, students worked through the variety of texts and "treatises" of the *Library of Useful Knowledge.* The selections included works by Lardner, Kater, Potter, and Young (Mechanics, Hydrodynamics, Pneumatics, Steam Engine), Brewster and Jackson (Optics), Herschel, Gummere, and Norton (Astronomy), Lyell, Trimmer, De la Beche, Bakewell, Allen, Philip, Dana, and Ansted (Geology and Mineralogy); Golding Bird, Muller, and Peschel (Physics); and Agassiz and Gould (Zoology). Whatever the themes or texts, Rogers distinguished himself from his predecessor at Virginia by incorporating the useful arts approach. Students, for example, received more instruction in the useful, or "mechanic," arts and expanded sections on theories about "dynamics" and "heat" after he arrived there. By the same approach Rogers expanded the options available to students in geology and mineralogy as part of his course in natural philosophy. He taught classes on the concepts of stratigraphy as well as "practical and descriptive portions of the Science." For mineralogy he went a step further, including "an economic view."[9]

Making further use of the elective system at Virginia, Rogers pressed for the establishment of a new school to emphasize his useful arts ideal. A year after joining the faculty, he led the founding of the first "School of Engineering" at the university. Upon receiving approval from the Board of Visitors, Rogers and a colleague, mathematics professor Charles Bonnycastle, created the school with practice and theory in mind. They divided six themes between them. Rogers taught three options: Geology, Heat and Steam, and Theoretical Mechanics. Bonnycastle took on the other three: Graphical Mathematics, Surveying, and "Theory of Roads, Railroads, Canals, Bridges." Rounding out the schools offerings, the two professors co-taught an additional surveying section as well as a section on "drawing and sketching." As for laboratory experiences, Rogers provided students with exercises "taught practically in the field" for practice in surveying.[10]

Toward the end of his career at Virginia, the institution allowed Rogers to reorganize the school of natural philosophy. He divided the area into two parts, junior and senior classes. The division closely matched his belief that students should first learn the principles and theories of science before attempting to apply such knowledge for practical purposes. Juniors received lectures on the theories of equilibrium, resistance, acoustics, and electricity, among other themes. Senior students, having a background in theory, explored "practical statics" and "practical dynamics" as related to architecture, steam engines, and additional practical topics. Rogers's emphasis on having students learn theory before proceeding to practice would become a hallmark of his educational reform efforts.[11]

Although Rogers enjoyed some success in advancing curricular changes, during his tenure at William and Mary and the University of Virginia, he gained virtually no ground toward his most prized ideal: laboratory instruction. Colleges of the era rarely strayed from such traditional methods of instruction as the recitation, a practice that required students to memorize texts for classroom recitals. Indeed, when the University of Virginia adopted lectures and written examinations, many viewed the methods as progressive or radical when compared to recitations. But even at Virginia, at best Rogers could only include laboratory demonstrations as part of a lecture. Otherwise known as "lecture demonstrations," the practice called for a faculty member to perform an experiment for students to observe. Rogers believed, however, that recitations, lectures, written exams, and demonstrations paled in comparison to work done in the laboratory, which allowed students to experiment with connections between theory and practice that no other form of instruction could provide. The University of Virginia nevertheless resisted this part of his reform efforts.[12]

The resistance that Rogers and science promoters at other colleges faced was due, in large measure, to a landmark publication in 1828. That year Yale College, one of the most influential institutions of the period, dampened enthusiasm for practical and scientific education by issuing a position statement, "Original Papers in Relation to a Course of Liberal Education." The expansion of natural science offerings had raised questions about the meaning of a classical education, questions that the article sought to put to rest. In effect, the report advanced an ideal of higher learning that placed the classics and mental discipline at the center of the American college curriculum.[13]

The policy-shaping document, known as the Yale Report of 1828, had two principal authors. Yale president Jeremiah Day drafted the first of two sections, which defined the aims of a liberal education: "Its object is to lay the foundation of a superior education." Day distinguished the universal applicability of the collegiate education from the particularistic training of practical studies. James Kingsley, who served as professor at Yale for a half-century, wrote the second section to explain the continued use of the so-called dead languages, primarily Greek and Latin. He described a need for the ancient languages in cultivating students for the literary world, gaining taste, strengthening the mental faculties, and preparing for the professions. Bringing together the perspectives of a president and a faculty member at Yale in

1828, the report outlined a tradition and philosophy of American higher education that had at one time been assumed. Both sections supported the notion of faculty psychology that depicted the mind as having "discipline" and "furniture" to acquire. Discipline consisted of exercising the mind, conceived of as a muscle, through recitation and rote memorization. The best furniture, argued Day and Kingsley, came from reciting the classical languages. Faculty psychology, pervasive throughout the Yale Report, influenced the standard conception of a proper collegiate education. All other forms of discipline and furniture, such as experimentation and scientific knowledge, asserted the article, had complementary, if lesser, roles to play in the undergraduate course of study.[14]

For Rogers and other science advocates within traditional colleges, the report left many educational questions unanswered. Why did the classics provide the best furnishings for the mind? Why should science faculty teach by way of recitations, the methods of language instructors, when the laboratory offered more for scientific studies? Why should professional and practical education be marginalized and not taught with the same rigor and given the same value as the classical curriculum? Ultimately, Rogers kept questions of this sort in mind as he plotted his college reforms.[15]

Rogers first expressed his vision of a technological institute in print during his tenure at the University of Virginia. After notice of his reform interests reached several members of Philadelphia's Franklin Institute, in 1837, the managers of the institute requested his assistance in developing a memorial for a "School of Arts" with which to petition the Pennsylvania legislature. For this concept of "school" Rogers drafted a proposal aimed at "professional education" for the mechanic arts. He modeled the program after medical and legal training. Young mechanics, he contended, worked in the nation's newly expanding fields of engineering and mining and manufacturing, almost always without formal preparation. The memorial offered an opportunity to establish a program to meet the needs of mechanics; central to his proposal was the notion of offering a greater number of theoretical studies than what was available at vocational institutions and more practical experiences than traditional institutions of higher education.[16]

The memorial recommended starting a school divided into six departments, each led by a head professor who was assisted by "sub-professors," or "practical instructors." Faculty had the option of using the recitation method and giving lectures, but the laboratory, Rogers emphasized, would provide the primary mode of instruction. In three departments—Mechanical Engineering, Chemistry, and Mathematics—the laboratory consisted of facilities with apparatus for instruction. The mechanic arts students would apply the principles learned in lectures to "model-making" through experimentation with "machinery" and "structures." Chemistry students would perform analyses of soils, minerals, and other substances under the guidance of specialized instructors. Those in the mathematics department would apply "principles of perspective and descriptive geometry" to architectural, topographical, and machine drawing. In the three remaining departments—Geology, Civil Engineering, and Agriculture—the outdoor environment itself provided the laboratory. Lessons in the field, Rogers argued, taught observational and operational skills required for enlightened mining, surveying, and farming.[17]

The proposal attributed several advantages to laboratory work. For one thing it exposed students to equipment commonly used in practice. Most colleges of the period avoided what Rogers's plan set out to do, given the high costs of the apparatuses of the day. Those institutions that did purchase such equipment considered them too valuable for student use. Instead, faculty limited apparatus use to lecture demonstrations or displays for campus visitors. Rogers's proposal challenged conventional wisdom by seeking to place the tools of science in student hands. For Rogers the laboratory also offered students an opportunity to translate theory into practice. Whether in geology, chemistry, or physics, he firmly believed, few means outside of laboratory instruction afforded such a connection. By experimenting with apparatus or in the field, students could observe phenomena and then describe their observations "to an exact form on paper." By controlling and describing their own experiments, students could thus become directly part of the process for learning new principles and theories. At the same time, laboratories had the potential to demystify occupations traditionally shrouded in folk myths or superstitions. Mining, for instance, "would be directed by sure principles and not by blind chance, or by a routine more often inapplicable than appropriate." "To be an enlightened mechanic," he continued, "it is also necessary, to a certain extent, to be acquainted with science; nor is it less true, that a knowledge of some of the arts is requisite to the cultivation of science itself." Rogers's plan offered a "more intimate union" of mind and hand.[18]

In justifying the new model of higher learning, Rogers reasoned that such a union between theory and practice had not always been necessary. In the "early stage" of the arts, before the industrial revolution, artisans gained sufficient skills from informal settings. Rogers referred to an elaborate system of apprenticeship in which trade workers began as journeymen under the direction of masters. Apprentices often aspired to reach the status of master, achieving proficiency enough to employ and teach other journeymen. With the onset of the industrial revolution, however, factories and other systematized forms of labor organization brought the apprenticeship system to an end. Trade workers of the day no longer acquired skills that would directly advance them in their field and within their community. Instead, they performed routine tasks, resulting in fewer opportunities for social mobility. Rogers's proposal for a School of Arts responded to a growing desire in the United States for a "professional" mechanic. By the same token, he reasoned that the school would not only benefit the mechanical, manufacturing, and agricultural communities but also the state more broadly. Writing in an era of internal improvements, he remarked that the "succession of experiments, often blindly undertaken" in areas such as surveying and geological studies had resulted in "disaster" and great cost. By providing a new education for a new class of mechanics, the state would profit through "the widest possible diffusion of that accurate and enlarged practical knowledge; for the want of which, labor and money are so often fruitlessly and perniciously expended." Rogers thus aspired beyond curricular and pedagogical reform, envisaging a form of higher learning that had consequences for social reform as well as for the advancement of science.[19]

In the end the school, with its six departments, laboratory method, and ambitions for social and scientific advance, never opened. Financial panic, similar to that of 1819, returned in 1837. The Pennsylvania legislature looked for ways to reduce or reject additional expenditures, and Rogers's proposal had little chance of surviving the retrenchment. Although the plan received support from the Franklin Institute and other quarters, the economic turmoil and reluctant legislature withheld the funds necessary for establishing the new school. Within a year the idea had all but disappeared.[20]

❧❧

Throughout the 1840s Rogers kept abreast of efforts to advance practical and scientific studies as they continued to emerge in the Northeast and elsewhere. Most colleges by 1840 had a professor of mathematics who, unlike their predecessors, taught the subject without also teaching the sciences or the classics. Joining their ranks, natural philosophers enjoyed somewhat better facilities; their courses also appeared more regularly as diverse offerings in optics, electricity, meteorology, and astronomy. Although student laboratories remained rare, new scientific equipment made the lecture demonstration more common. Recitation, however, continued as the dominant practice, even among science instructors who occasionally used alternative teaching methods. Chemistry joined mathematics and natural philosophy in making inroads into the curriculum. The popularity of chemistry courses stemmed from the medical, industrial, and agricultural applications of the field. Chemists often taught mineralogy, geology, and agricultural topics as a single course. Yet as research in each of the areas became more specialized, chemists began to expand their course offerings to match the trend.[21]

The college curriculum grew significantly during the 1830s and 1840s, the period in which Rogers labored over his ideal for higher learning. The expansion of the curriculum and only slight changes to graduation requirements created obvious challenges that antebellum scholars had to face. Collegiate leaders, many of whom continued to look to the Yale Report of 1828 for guidance, attempted to address the "crowding of the curriculum." Some considered lengthening their undergraduate programs from four to five or six years. Others experimented with certificates or science diplomas. Still others looked to nondegree or partial studies programs to meet demands for science offerings. Each of these schemes were attempts at the same thing: to preserve the integrity of the classical curriculum (and the bachelor of arts degree) and offer science on the side. While at the University of Virginia, Rogers benefited from a system of electives that allowed students to take his science courses as part of their program of study. The university had promoted this student freedom for decades, and over time other institutions began to take notice. But observers found the structure of Virginia's independent schools and system of electives difficult to adapt to traditional programs. Although an endless accretion of courses, parallel and partial programs, and electives failed to satisfy reformers, a popular solution soon arose from Harvard and Yale.[22]

By 1847, a decade after Rogers wrote the School of Arts proposal, both Harvard and Yale had established plans that led to separate schools of science and practical studies. At Harvard the school focused on "engineering, mining, mechanical drawing, and methods of constructing machinery" but later, under Agassiz, shied away from applied science. The Yale plan, meanwhile, kept a more applied tone, but received no financial support from the college. Faculty there relied exclusively on student fees. For some Harvard and Yale provided a model that was attractive; it kept practical and scientific education outside of the traditional curriculum, thereby reducing interference with an institution's established mission. But this reform, like others of the period, left many others disappointed. Charles W. Eliot pointed to the absence of entrance requirements as one source of the problem. "Anybody, no matter how ignorant," could find a seat at science schools such as those of Harvard or Yale. At Harvard another problem arose from the lack of adequate facilities and laboratory experiences for students. Eliot later recalled his experiences at the Lawrence Scientific School and its "humble" beginnings. The faculty had had no means of "offering laboratory practice to the students, except as a favor which could be granted to very few."[23]

For Rogers the reforms of the era seemed ineffective, whether in the form of increased requirements or alternative courses of study or separate schools. The continued lack of laboratory experiences for undergraduate students at Harvard, at Yale, and at most every other program in the country, he argued, continued to pose an obstacle to scientific progress. He saw a need to break new educational ground. "The Lawrence School," observed Rogers at the time of its founding, "never can succeed on its present plan in accomplishment of what was intended. It can only, as now organized, draw a small number of the body of students aside from the usual college routine." The heart of the matter for Rogers rested, as it often did, in the imbalance between theory and practice. Harvard's science program "should be *in reality* a school of *applied science*, embracing at least four professorships, and it ought to be in great measure independent of the other departments of Harvard." He believed that only a truly independent program would have enough freedom to develop studies along the lines of the useful arts, to make science more than an "aside" in the curriculum, and to offer alternative modes of instruction. Such a program, Rogers continued, would "embrace experimental physics," for example, to expose students to "applied mechanics" and the "principles" underlying such applications.[24]

Not surprisingly, Rogers took a special interest in the organization of Harvard's Scientific School, for in 1846, the year before its founding, he had sent an elaborate proposal for an institute to reformers in the Boston area. In a letter sent through his brother Henry, who had moved to Boston, William described to John A. Lowell Jr., director of the Lowell Institute, the potential for starting a useful arts—based school of science. William's proposal to Lowell argued for scientific studies that incorporated advanced laboratory instruction.[25]

The preface to Rogers's proposal defined the main objects of a polytechnic school. He contended that, foremost, an institute should promote "the inculcation of all the scientific principles which form the basis and explanation of them, and along with this a full and methodical review of all their leading processes and operation in connection with physical laws." The basic principles, as he envisioned them, placed the student "in the workshop" for clarity of comprehension of "the agencies of the materials and instruments with which [the student] works." As in his previous plans for reform, he wrote that by such practical and applied instruction through laboratory work, the student "is saved from the disasters of blind experiment." The document argued that experimentation in the laboratory would prepare a student for practice in the field. From Rogers's own experience managing field assistants for the Virginia survey, "blind experiment" produced fruitless results. He dismissed the increasingly common use in the United States of the lecture demonstration as inadequate for his alternative model of higher learning. According to the proposal, an institute should provide an advanced education for the mind and thorough training of the hand.[26]

Rogers continued his description of the ideal science institution by outlining its organizational structure. His plan divided the model program into two areas, one for theory and the other for practice. The first department of the school, guided by two instructors, would focus on the "groundwork of . . . general physical laws." Matriculants in this department would receive a broad introduction to general principles of physics. Students, he warned, could not benefit from the second division until they had successfully mastered the elements of instruction provided in the first department. The second division would have an "entirely practical" emphasis, in which students would learn "chemical manipulation and the analysis of chemical products . . . elementary mathematics . . . [and] full instruction in drawing and modeling." The laboratory, meanwhile, stood at the center of student experiences, with "two or three tutors, or sub-professors, to give personal instruction in the laboratory." Rogers's plan had far-reaching ambitions; he expected that soon the school would "overtop the universities of the land in the accuracy and the extent of its teachings in all branches of positive knowledge."[27]

Rogers followed with a series of examples describing how practitioners such as machinists, engineers, and architects could benefit from theoretical and practical instruction as offered by his proposal. These practitioners, he insisted, needed more than a passing acquaintance with physical laws such as the dynamics of equilibrium, friction, resistance, chemical and thermal changes, and mechanical principles. Rogers predicted that refineries and manufacturing would soon require the service of those well versed in theory and scientific laws. "The processes they [refining and manufacturing] involve," he suggested, "are but the vast practical enlargement of the common experiments of the laboratory and lecture-room." He emphasized the point, one grounded in the useful arts ideal, by asserting that "there is no branch of practical industry . . . which is not capable of being better practised, and even being improved in its processes, through the knowledge of its connections with physical truths and laws, and therefore we would add that there is no class of operatives to whom the teaching of science may not become of direct and substantial utility and material usefulness." The proposal made clear that the "operatives" he had in mind were composed of a diverse audience of science enthusiasts. He wanted to draw "lovers of knowledge of both sexes to the halls of the Institute." His support of coeducation distinguished him from many of his colleagues. At the University of Virginia, from which Rogers wrote to Lowell, similar coeducational ideals would not appear in practice for more than one hundred years.[28]

Although Rogers offered an elaborate argument for an institute based on his years of research and administrative experience, Lowell resolved to do nothing with the proposal. The funding source that supported the Lowell Institute—Lowell Sr.'s will—and the ideals of the governing body directing the institute likely prohibited the use of funds for such a project. Yet Rogers's 1846 plan had indirect as well as direct influences on selected educational reforms of the 1840s.

The proposal likely had an indirect influence on the founding of Harvard's Lawrence Scientific School. Abbot Lawrence, the Boston industrialist who donated fifty thousand dollars for the science school, was no stranger to John A. Lowell. They worked together on several projects between March 1846, when Lowell received Rogers's plan, and June 1847, when Lawrence notified Harvard of his donation and the accompanying stipulations. Lawrence and Lowell shared executive duties in a textile manufacturing plant and were on a local committee for the 1847 meeting of the Association of American Geologists and Naturalists. Lowell, moreover, held a seat as a fellow of Harvard College at the time Lawrence was proposing his bequest. The two industrialists thus had ample opportunities across the fifteen-month period to discuss proposals for a science institute. Still more revealing, however, was the similarity between the Rogers and Lawrence plans. Rogers had called for two professorships to cover the fields of physics, chemistry, and geology. A similar structure appeared in Lawrence's plan, in which he sought to create professorships for the fields of engineering, chemistry, and geology. That Lawrence emphasized a geology chair with specialization in the "industrial arts" further supports the view that Rogers's ideas may have been incorporated into the Harvard program. While many other proposals could have influenced Lawrence's scientific school, there were significant, if indirect, elements shared with Rogers's plan.[29]

The practical, or applied, mission tied to the Lawrence donation came undone, however, at the hands of Louis Agassiz, who diverted the school toward basic, abstract ends against the founder's intent. The ensuing tension may have led Agassiz to press for adding Rogers, who promoted the useful arts ideal, to the Harvard faculty. "He [Agassiz] told me in confidence," confessed a colleague to Rogers, "of his wish and purpose to make room for you in the scientific school or new museum as professor of Geology, he wishing to relinquish it and retain the Zoology, as soon as the museum affairs are organized. If this should suit you, I do sincerely trust it will be done." By the time President Cornelius C. Felton and members of the faculty at Harvard began to consider Rogers for an appointment, who was then unaffiliated with an institution, he'd decided that it would be better to remain independent. Felton, noted Rogers, "expresses the strong wish of himself and others to have me in Cambridge. They are proposing to establish a Professorship of Geology and Mining in connection with the Lawrence School, at least for the present. . . . If I enter Cambridge I can do so without in the slightest degree relinquishing the individuality I have heretofore maintained. So, at least, I think, and on no other terms would I be willing to connect myself with the college." Freedom to actually practice the useful arts he had been preaching, Rogers concluded, wouldn't come easy at the tradition-laden institution. In the end an appointment at Harvard never materialized for Rogers, but it was hardly his last interaction with the institution.[30]

Rogers's 1846 plan also had a direct influence on fellow reformer Francis Wayland, president of Brown University. Wayland became well-known as a staunch critic of the classical college, the most vocal and popular critic to emerge after the publication of the Yale Report of 1828. In the 1840s and 1850s he advocated radical changes to the traditional undergraduate course of study in light of an emerging middle class. As a political economist, Wayland analyzed the condition of American colleges and concluded that because they offered superficial, rigid, and antiquated courses, they stood on the brink of bankruptcy. Protesting the university's resistance to change, Wayland went so far as to resign the presidency of Brown, but after negotiating with the administration, he retracted his resignation on the condition that the Brown corporation willingly undertake a wholesale revision of its program. The trustees agreed, raised $125,000 for the new curriculum at Brown, and welcomed Wayland back to campus. Having gained popular support for his ideas and a vote of confidence from the institution's governing board, he looked to his contemporaries for model reform plans.[31]

One of the first educational leaders that Wayland went to for inspiration was Rogers. After William's brother Henry met with Wayland in December 1849 about a proposed restructuring of Brown's curriculum, Wayland appealed to William for specific ideas concerning changes to science instruction in American higher learning. Henry assured William that "Wayland is intent upon some valuable and important collegiate reforms, and his views are shared by [Zachariah] Allen [a manufacturer and trustee of Brown] and a majority of trustees." "They contemplate an entire reorganization of their college," he continued, "introducing much more science and practical instruction, less Greek, etc., and adapting some of your system. Wayland is tired of the old monastic system . . . [and] wishes a copy of your exposition of the system, etc., at the University, Memorial to the Legislature, and any documents or notes of your own having a bearing on the subject. He has had a copy and lent it to some of his trustees, and it may not suffice for his wants just now, therefore send him another."[32] Wayland based his well-known *Report of 1850* in part on Rogers's ideas about reform. Among the most controversial of its elements were an end to the fixed curriculum, the beginning of a free elective system, and the establishment of a robust program of applied science, in addition to courses in agriculture, law, and education. At the center of the *Report* stood many of the useful arts principles that Rogers had advocated for almost a quarter-century, principles that called for the union of theory and practice and for opening up the curriculum to applied instruction.

In support of his *Report,* Wayland traveled to Virginia in April 1850 to discuss reform with Rogers in person and to review the system of electives on the Charlottesville campus. William hosted Wayland's stay, as they conversed about reform, toured lectures on the campus, and met with other faculty. In his memoirs the Rhode Island visitor recorded that he traveled to the institution "wishing to gain all possible aid from the light of experience." Following the visit Rogers wrote to his brother Henry that he "appear[s] quite determined to adopt our more liberal features in their [Brown's] new scheme." Although William's useful arts emphasis appeared in Wayland's reforms, Brown's approach to the new course of study and supply of funds proved inadequate for any long-standing reform. Five years after the publication of the *Report,* Brown trustees ousted Wayland and returned the institution to the original design.[33]

While many factors contributed to the decline of this reform effort, Wayland's lack of scientific research and engagement with the profession stands out among them. As Rogers would soon discover, credibility in the sciences when advocating scientific and practical studies came to play an increasingly important role in American higher educational reform movements of the mid-nineteenth century.

CHAPTER SIX

Instituting a New Education

A LTHOUGH ROGERS encountered setbacks to his first two proposals, those in Philadelphia and Boston, the experiences became rehearsals for his third reform effort. Across the higher educational landscape of the 1850s could be seen institutions that satisfied parts of his own plan, but Rogers sought to bring these scattered innovations together. He advocated a vision for a "comprehensive" institute, meaning an institute that offered students several specialties within engineering, provided opportunities to specialize in the natural sciences, and taught by way of laboratory research. His vision also placed these elements within the useful arts framework. For him existing programs erred either on the side of practice or theory. He believed that no institution had provided a comprehensive curriculum or had struck a balance guided by the useful arts.[1]

West Point came close. Founded in 1802, the military academy offered the first engineering program in the United States. By the 1830s the academy had become well-known for preparing engineers and scientists for practical fieldwork. Francis Wayland, during his reform years at Brown, declared that "the single academy at West Point has done more toward the construction of railroads than all our . . . colleges united." William's brother Robert Empie Rogers had a similar impression when he stated that "engineering holds but very few inducements, for only those who have been educated at West Point stand in the way of promotion. . . . They alone are sure of constant occupation." Through the efforts of superintendent Sylvanus Thayer, the curriculum borrowed heavily from the French military educational system of the Ecole Polytechnique. Both Thayer and later his colleague Dennis Hart Mahan visited France for training in engineering and to review the latest literature in the field. For most of the antebellum period the military academy offered a four-year curriculum based on their efforts. Nearly three-quarters of the course work focused on civil and military engineering, mathematics, and natural philosophy. Interspersed within these areas was training in "military tactics" and the "science of war," geared toward preparing cadets for combat.[2]

Two basic obstacles hindered the program from establishing the kind of institute

Rogers advocated. The first problem had to do with the emphasis on producing military officers. As a military, rather than an academic, institute, West Point's elementary admissions requirements of basic reading, writing, arithmetic, and physical health were established and controlled by Congress. Contemporaries blamed the minimal admission requirements for difficulties with the student body. Second, between 1855 and the Civil War the course of study turned almost entirely practical. With a curriculum composed of military drilling and field skills practice, theory and research virtually disappeared from the daily routine of students.[3]

The Rensselaer School, later a polytechnic institute, also came close but in a manner somewhat different than West Point. New York state political leader Stephen Van Rensselaer founded the school in 1824 in order "to qualify teachers for instructing the sons and daughters of farmers and mechanics, by lectures and otherwise." As a teacher training institute, the program placed a strong emphasis on applied topics, leaving preparation for research at the margins. One of the program's strongest areas was its use of the laboratory for instruction. As such, under the leadership of geologist Amos Eaton, the school earned the distinction of being one of the first institutions to provide laboratory experience for its students. Meanwhile, the curriculum lagged behind its innovative instruction. As late as 1848, beginning students could complete the course of study in one year, and advanced students finished in a mere twenty-four weeks. By the 1850s the curriculum had been reorganized, with the offerings pared down to only architecture and engineering and the requirements raised to three years of study.[4]

Edging substantially closer to what Rogers had in mind, Rensselaer's Benjamin Greene proposed a plan for a "true polytechnic" in 1855. Inspired by the German system of science instruction, Greene published a proposal to transform the school into a comprehensive institute. Rather than having another school for architecture and civil engineering, he argued that the United States was ready for an institute that offered degrees for aspiring mining engineers, mechanical engineers, and practical chemists. Furthermore, he described a true polytechnic as offering broad training for the mind and the body. To develop the mind, Greene suggested courses in the social sciences, law, ethics, aesthetics, landscape design, and literature. He envisioned approximately twenty-six faculty members assigned to these and other subjects. For the body he described a thorough regimen of gymnastics. Physical fitness provided "presence of mind, consciousness of physical capacity, power of command, and promptness of action," all of which were required for the safe conduct of fieldwork. In the end, however, Greene resigned over an embezzlement scandal four years after publishing his proposal. His plan fizzled, leaving unchanged the much narrower architecture and engineering program already in place.[5]

Other reformers of the 1850s groped for a handle on the public's desire for science and technology, but many of their ideas fell on hard times or unfortunate circumstances. Henry Tappan promoted the idea of graduate-level science instruction at the University of Michigan, but he experienced difficulty after a tenuous start. During his tenure only a handful of students enrolled in his programs, which unlike those at West Point and the Rensselaer School, emphasized theory over practice. Cornell's president Andrew D. White later dubbed the practical offerings at Ann Arbor as "wretchedly meagre." Similarly, James Hall, A. D. Bache, and the Lazzaroni backed the idea of a university in Albany, New York. Focused on advancing theoretical science, the plan sought to bring together the Lazzaroni in one place to create a center for scientific research and teaching. Economic and political setbacks ultimately torpedoed their effort. Other short-lived and long-lasting institutions such as Norwich (1820), the Gardiner Lyceum (1823), the Franklin Institute (1824), the Citadel (1843), the U.S. Naval Academy (1845), the Polytechnic College of Pennsylvania (1853), the Brooklyn Polytechnic Institute (1855), Cooper Union (1859), and schools in Cleveland, Ohio (1857), and Glenmore, New York (1859), showed educational reform activity but little resembling Rogers's comprehensive or useful arts aspirations.[6]

Looking beyond New York, Michigan, and Massachusetts, Rogers followed the educational developments occurring in Europe. Some have suggested that his main European influences came from German institutions, such as the Karlsruhe technical academy, but it is clear that he drew on many different models and systems that supported his useful arts ideal. He was eclectic in his approach when examining the "greatly received . . . Polytechnic and Scientific schools of Carlsruhe [sic] and Zurich, the Ecole Central, School of Mines, and the Polytechnic school of Paris." Of all of these influences he took a special interest in the Conservatoire des Arts et Metiers and the Ecole Central des Arts et Manufactures, both located in Paris. The Conservatoire, founded during the French Revolution, established a model for the modern industrial museum. Its exhibits featured products of applied science, in addition to lectures from distinguished scientists, a technical library, and laboratories for research. The museum collected "machines, models, tools, drawings, descriptions, and books in all the . . . arts and sciences." These collections benefited from a government mandate requiring that "the originals of instruments and machines invented and perfected shall be deposited at the Conservatoire." Rogers wanted a similar museum to display American industrial and agricultural innovations for the purpose of higher learning, the advancement of science and technology, and the general diffusion of knowledge.[7]

Rogers also believed that the United States needed an institute similar to the Ecole Central. This is "what is wanted for American students," for "the students at the Central School of Arts and Science in Paris," he observed, received a "broad foundation of scientific study, and building upon it practical education." Established in 1829, the Ecole Central was built upon the goal of offering a civilian alternative to the country's prestigious military engineering program, the Ecole Polytechnique. The civilian program, unlike the military school, provided students with an elective system that allowed for specialization in diverse areas of civil and industrial engineering. Graduates from the Ecole Central prepared for such fields as agriculture, architecture, railroad engineering, textile manufacturing, public works, industrial chemistry, general civil engineering, machine manufacturing, metallurgy and mining, and commerce. In response to the largely mathematical and theoretical training of the Ecole Polytechnique, the civilian program balanced theoretical training in geology, physics, and chemistry with practical laboratory exercises and the workshop. A majority of students in these laboratories and workshops came from the business, industrial, and laboring classes. As the first private engineering institution, its founders effectively ended the association of engineering with the functions of the state and military, attracting students interested in preparations perceived as necessary for the management of French industrialization. The useful arts stood as the Ecole Central's most important, if unstated, principle, one that Rogers recognized and sought to import to the United States.[8]

By the time Rogers arrived in Boston in 1853 as an independent scholar, he had studied carefully what he considered flaws in his own country's leading schools of science. "I learn that students are greatly pleased," chided Rogers about Harvard's chemistry program, "because, for the first time, they are shown some chemical experiments. Last year they committed the chemistry to memory!" Thus, as late as the 1850s, Harvard was still using recitations and had only begun to experiment with lecture demonstrations.[9]

Six years after arriving in Boston, a sequence of seemingly unrelated events began to unfold that brought Rogers closer than ever to his final educational proposal. That January, Massachusetts governor Nathaniel P. Banks gave an address that discussed the need to incorporate "educational improvements" in Boston's Back Bay policy. The Back Bay, according to an observer of the period, consisted of "a broad shallow basin of salt water lying to the westward of the narrow peninsula upon which the city then stood." State leaders had commissioned the draining and filling of the basin to create "new land," the heart of Back Bay policy. While the primary purpose was to alleviate the congested areas of Boston proper, Banks tied an educational initiative to the project. He suggested to the legislature that it "make provision for the application of this property to such public educational improvements as will keep the name of the Commonwealth forever green in the memory of her children; and to this end I earnestly recommend . . . that the first public charge to be made upon this property shall be for the enlargement of the public-school fund." In addition to the school fund, Banks talked vaguely about supporting "efforts with the cooperative power of individuals, associations, and institutions, partially or altogether devoted to science." Almost as soon as the idea was mentioned, two proposals came before the legislature.[10]

One of the plans came from advocates of the public school fund. George Boutwell, secretary of the State Board of Education and a staunch supporter of public education, led the effort. For years he had argued for the need to increase teacher salaries, the school fund being one promising means to such an end. As a former governor of Massachusetts, Boutwell also understood how to navigate a funding proposal through the halls of the state legislature. Not surprisingly, the plan easily passed into law in March 1859 requiring that "the proceeds of all sales of the . . . Back Bay lands . . . shall be added to the School Fund, until the principal of said fund shall amount to the sum of three million dollars." Considering that a year earlier Boutwell had asked for an increase of a million and a half dollars, which he called "desirable, if not necessary," the new legislation succeeded in advancing the public school cause.[11]

The other proposal, which Rogers supported, was not so fortunate. It expressed the interests of Boston area societies seeking to petition the state collectively for Back Bay land. The societies, including those of "Agriculture, Horticulture, Art, Science, and various Industrial, Educational, and Moral Interests of the State," elected Samuel Kneeland Jr. to represent them before the legislature. Kneeland, taking the lead in drafting the memorial, requested four squares of Back Bay land on which to build what they planned to call the Massachusetts Conservatory of Art and Science. This conservatory, he argued, would facilitate interaction between the disparate groups represented in the proposal, reducing duplication in their collections and making their holdings accessible to citizens of the state in a single location. Kneeland and a conservatory advocate on the legislature, Thomas Rice Jr., described briefly how the four squares of land would correspond to four major interests in the conservatory. One square would be "devoted to collections of Implements, Models, and other Objects pertaining to Agriculture, Horticulture and Pomology." In the "halls and grounds" of this section, advances in agricultural machinery as well as practices would be "scientifically explained" to the public. Another square would house museum specimens of "Natural History, Practical Geology, and Chemistry" related to the state. The Boston Society of Natural History, among the memorialists overcrowded with specimens, stood to benefit from the spacious Back Bay lands. A third square corresponded with "the development of Mechanics, Manufactures and Commerce." Offering general education to the public, this section would display the latest technologies of "this age of invention," technologies related to all aspects of industry. For the final square the document assigned a museum of "Fine Arts, History, and Ethnology."[12]

Beyond writing his signature on the Kneeland document, Rogers played virtually no role in preparing the proposal. At the time it came before the legislature, he was on a lecture tour in Virginia. When he returned to Boston, he learned that state leaders had sent mixed messages. A report by a Joint Special Committee, on the one hand, acknowledged that a Conservatory of Art and Science would benefit citizens of the state. In particular, the committee appreciated the educational function of the museum, one that "would not only add to the material prosperity of our own State, but by drawing strangers from all parts of the country would become the means of diffusing knowledge to an extent which can hardly be estimated at the present time." The report captured the spirit of the proposal, declaring that "this practical age demands a practical as well as theoretical education." On the other hand, because of the number of bills before the assembly and the "lateness" in the session, the same committee concluded that "the present is not a propitious time for action." Although the vote against the plan ended Kneeland's effort, Rogers took note of the items of interest to the legislature.[13]

Mere weeks after Kneeland's defeat, the societies that had attempted to organize a conservatory turned to Rogers, asking him to prepare a second proposal on their behalf, one that would build on Kneeland's work. Swift to act, Rogers had a new proposal ready in time for the January session of the 1860 State Assembly. For this assignment he faced an important challenge: finding a focus. The range of interests represented, from art to agriculture to industry to natural history, worked against developing a single, persuasive argument for the project. Another challenge had to do with internal politics. Some societies already had established facilities elsewhere and sought to expand on the Back Bay; others existed in name only, waiting for their first square of land on which to build. Rogers's plan for the conservatory attempted to mediate between interested parties by using the same four divisions presented in the previous document, only with fuller elaboration on each point.[14]

Two changes from the earlier document, however, clearly show Rogers's own designs and offer clues about his intentions for the proposal. Rogers first recommended a laboratory. In doing so, he placed great emphasis on the social function of scientific research. The laboratory, based in the Agriculture division, would produce "research as demanded by the progressive and scientific husbandry of the present day . . . Associated with such a museum, we should look for the organization of a laboratory equipped for every branch of chemico-agricultural experiment, which, while furnishing reliable reports on the composition of soils, manures and vegetable products, and thereby protecting the agricultural public from the impositions so frequently practised by dishonest or ignorant pretenders, might by its larger researches help to advance the theory as well as the practice of agricultural processes."[15] For many Americans the idea of such a laboratory would have sounded unfamiliar. Perhaps equally foreign to readers, Rogers also added a system of useful arts instruction to the project. He described without much elaboration his ideal of a "comprehensive polytechnic college . . . a complete system of industrial education." While rendered here in abbreviated form, Rogers hinted at his decades-old vision for an institute of technology. In all work, whether in museums, laboratories, or instruction, he emphasized the dual aims of abstract and practical knowledge.[16]

Segments of Rogers's plan written for the conservatory passed one body of the assembly. The House of Representatives favored allowing only two of the four divisions of the plan to be considered for review by the Senate. The Boston Society of Natural History and the Massachusetts Horticultural Society gained approval, representing the geological and agricultural divisions, yet the proposed industry and fine arts sections were left behind. House members favored the two they passed because the institutions were established, already in operation, and ready to build on the Back Bay squares. The other two would take longer to translate from ideas into institutions, all the while affecting land values, the sales of which went toward the school fund. Despite expected improvements to land values, the Senate voted against the entire plan, delivering a second defeat to conservatory supporters. [17]

The reasons the proposals were defeated are not entirely clear. Both proposals suffered from too broad a scope, Rogers's more so than Kneeland's. Had the two men limited their projects to specific goals, rather than extending them to a consortium of museums, they might have been more successful. The state had already approved generous support to Agassiz for the founding of a Museum of Comparative Zoology (MCZ) at Harvard. Funding for additional museums would have required a level of detail and distinction from the MCZ that neither plan provided. Kneeland, to his credit, had mentioned that Agassiz aimed at "the development of abstract science," while the conservatory sought to "cooperate with such labors . . . of a more practical character." Both memorials also referred to assisting those engaged in industrial occupations and to applying principles of purely scientific research. It would have been difficult for Agassiz to argue a similar position with the MCZ. Yet Kneeland and Rogers likely fell short of making necessary and clear distinctions such as those in Harvard's plan. The defeat may also be attributed to Bostonians' social and cultural perceptions of science at midcentury. During the 1850s and 1860s sponsoring education for the "industrial" classes didn't rank particularly high on the lists of priorities among Boston Brahmins. Louis Agassiz and the MCZ, both of them magnets that attracted generous support, represented established traditions in science, reverence for seeking a divine hand in nature, and abstract research. Rogers's ideals, meanwhile, represented the new industrial order, a defense of Darwin, and the useful arts. Agassiz experienced little difficulty when it came to fund raising; Rogers and his circle struggled to raise a mere pittance by comparison.[18]

Rogers's own speculation regarding the conservatory's downfall turned toward the school fund. He suspected, most of all, state leaders in George Boutwell's camp. Rogers believed many of them had succumbed to the idea that a gain for the conservatory meant a loss for the school fund. "After delays and reconsiderations," he lamented, "the Senate have finally refused to grant the Back Bay reservation for which we applied." From his experience working with state government officials, he concluded that "some enemies of the bill were quietly preoccupying the minds of senators, so that when the time for the action drew near we found that the narrow financial views instilled into them could not be corrected. Unluckily, the Back Bay lands were last year pledged to the increase of the common-school fund." Rogers had a point: if the fund derived revenue from the sale of Back Bay lands and the conservatory had requested a grant of four squares of land, then the fund would lose on the gift made by the state. Any Back Bay land given freely to the conservatory would deprive the schools of revenue.[19]

Whatever the reasons for the setbacks, Rogers gained some important political lessons. The legislature had shown an interest in the basic principles of the useful arts ideal. When the committee reviewing Kneeland's report declared that "this practical age demands a practical as well as theoretical education," Rogers knew he'd found potential supporters for his vision. He was also well aware of the opposition he would likely face. If Rogers took the lead in founding an institute of technology via the Back Bay lands, he would have to contend with the school fund advocates.

Armed with these political insights, Rogers began to prepare a third proposal, third in the sequence of his own reforms and third among those of the conservatory. This time, in the fall of 1860, he focused directly on his career-long ideal for higher learning. Working still with a committee formed by conservatory supporters, he began to advance his idea of creating a technological institute. As chair of the committee, he redirected the attention of his colleagues from museum exhibits to science instruction. For them he prepared the *Objects and Plan of an Institute of Technology, including a Society of Arts, a Museum of Arts, and a School of Industrial Science, Proposed to Be Established in Boston.* At a meeting of the conservatory he convened on

October 5, 1860, Rogers presented his document outlining a three-part plan for a Massachusetts Institute of Technology.[20]

The first part, a Society of Arts, proposed a research arm for the institute. Rogers defined the society as a "department of investigation and publication, intended to promote research in connection with industrial science." He used the term *industrial science* interchangeably with the *useful arts,* meaning an interaction between theory and practice in the sciences. Thus, when he referred to the duties of the society, he referred to its theoretical and practical aims. The society itself would consist of regular meetings held by members of the Boston area community interested in discussions, presentations, and the preparation of reports on the useful arts. Through oral and written communication this department would be responsible for keeping abreast of recent science-related "inventions, products, and processes," both domestic and foreign. Rogers described the need for establishing a new journal, the *Journal of Industrial Science and Art,* to bring together the communications. The periodical would record the proceedings of the society and the progress of MIT's museum and school and would provide "a faithful record of the advance of the Arts and Practical Sciences at home and abroad." Additional duties of the research branch of MIT included proposing and sponsoring research studies. Members would recommend experiments with products, processes, and machinery worthy of further investigation. If any innovation they tested had notable qualities, the society would present the maker with a special honor or award.[21]

To fulfill its mandates, Rogers suggested organizing twelve standing committees within the society of Arts. Some—such as the Mineral Materials, the Organic Materials, and Engineering and Architecture committees—dealt with products used in the useful arts. Others, such as the Tools and Implements, Machinery and Motive Powers committees, focused on practical machinery for agriculture and industry as well as "mathematical, chemical, and philosophical implements." Still others spanned the gamut of manufacturing processes related to textiles, wood, leather, and other products or covered a range of issues related to household economy (i.e., ventilation, the preparation and preservation of food, and the "protection of the public health"). While most of the committees specialized in products, machinery, or processes, the rest—committees on commerce and on the graphic and fine arts—incorporated all three.[22]

The second part of *Objects and Plan* involved the creation of a Museum of Industrial Art and Science. Like most museums, this one aimed at collecting and preserving objects of "prominent importance"—in this case, of importance to the useful arts. Rogers highlighted the need to arrange the specimens in a way that illustrated relationships between science and industry. The scientific relationships of mineral materials, for example, involved geological theories regarding the placement "on or beneath the earth's surface" of mineral formations. At the same time, illustrations of the composition and the means of extraction of such formations would provide valuable information to the "architect, engineer, and practical geologist, as well as those engaged in iron-making and other branches of metallurgy, and in the glass and ceramic manufactures." Organic materials, Rogers explained, would also be presented according to aspects of both science and industry. Indeed, such exhibits would illustrate "the whole history of each leading object, from its origin to its appropriation by the more advanced industrial processes." Information of this nature, he argued, aimed at the scientist as much as the artisan, manufacturer, and merchant.[23]

In addition to functioning as a repository, the museum would offer a global perspective. The exhibits would not only present homegrown innovations but also forge comparisons between ideas domestic and foreign. By comparing technologies of production and the production of technologies from around the world, Rogers hoped to "apprehend our relations to other producers." He had his sights set on a museum of industrial art and science that would save the artisan and manufacturer from blind experimentation.[24]

The School of Industrial Science and Art, the third part of Rogers's *Objects and Plan,* would become the most important of the three. This school, following the useful arts ideal, would encompass "systematic training in the applied sciences, which alone give to the industrial classes a sure mastery over the materials and processes with which they are concerned." By "systematic training" Rogers was referring to the organization of a technological institute for advanced science instruction. As he described it, the institute would offer many of the characteristics common to American higher education: recitations, lecture room teaching, examinations. Unique to Rogers's institute, however, was the unqualified centrality of "laboratory exercises." Faculty not only would demonstrate experiments as part of lecture presentations but also would supervise student experiments with laboratory apparatus. Through these supervised exercises, students were to acquire "fundamental principles, together with adequate practice in observation and experiment, and in the delineation of objects, processes, and machinery." The kind of students Rogers hoped to attract included those from the industrial classes. He praised the quality of public education in the Boston area, claiming that many graduates of the public system would benefit from further study at the proposed MIT. Rather than enter commerce, agriculture, or the mechanic and manufacturing arts without preparation, Rogers argued that the industrial classes could begin to enjoy professional status through the institute's scientific training. As the institute's resources grew, he hoped to bring "the entire systematic training of the School . . . within reach of aspiring students of humble means." More than mere popular lectures, Rogers aimed at the "highest grade" of scientific instruction in America for the useful arts.[25]

To Rogers's ambitious ends, the school would support five main departments: Design, Mathematics, Physics, Chemistry, and Geology. The Department of Design would teach drawing concepts and skills necessary to the work of engineers, architects, and machinists. Instruction in design, according to Rogers, brought together "geometrical, architectural, and free drawing, and the delineation of the apparatus and machinery of the arts." He imagined students working with fabrics and metals, producing a variety of figures, patterns, and models, and learning about "the principles of regulating the arrangement and combination of colors." These principles provided a "scientific basis and leading operations of the arts of engraving and photography." The Mathematics Department would also, for the most part, use alternative means of instruction. Introductory courses, from geometry to calculus, might start with traditional textbook and recitation methods, but students would then apply mathematical concepts in laboratory exercises involving surveying, navigation, and mapmaking, among other "constructive and manufacturing arts." Rogers likewise organized the departments of Physics, Chemistry, and Geology to incorporate theoretical as well as practical instruction. His *Objects and Plan* described a technological institute infused with the useful arts ideal.[26]

Student experiences with apparatus figured as the most important part of Rogers's school. His scheme "could not be prosecuted in a manner to be practically available without personal training in analysis and experiment, and would therefore demand the facilities of an ample and well appointed Laboratory." More so than other reforms in American higher education of the period, Rogers made explicit his emphasis on the laboratory for every student and every branch of science, for practice and theory. "The most truly practical education," he contended, "even in an industrial point of view, is one founded on a thorough knowledge of scientific laws and principles, and which unites with habits of close observation and exact reasoning . . . the highest grade of scientific culture would not be too high as a preparation for the labors of the mechanic and manufacturer."[27]

Members of the conservatory meeting, in which Rogers presented the *Objects and Plan,* immediately voted to support the idea for creating MIT. The conservatory adopted the proposal and approved its publication for distribution. Rogers had persuaded his colleagues that, because Europe had its Conservatoire des Arts, Ecole Central of Paris, Kensington Museum, the School of Mines, and Museums of Economic Geology and Botany, Boston needed the Massachusetts Institute of Technology.[28]

Having experienced defeat earlier in the year, Rogers had learned that his proposal required more than promising ideas. He first applied for an "Act of Incorporation for MIT," approximately one month after presenting the *Objects and Plan* to the conservatory. The Massachusetts secretary of state forwarded the application to the legislature scheduled to meet in January 1861. In the meantime Rogers made preparations to help ease his plan through the State Assembly by cultivating strong political allies. When *Objects and Plan* was printed and ready to be circulated, Rogers and his supporters prepared a list of a few hundred Boston area citizens whom they believed would take interest in his proposal. He mailed these supporters copies of the *Objects and Plan* accompanied by a circular that summarized his interest in founding MIT on the Back Bay lands, hoping that the idea, as expressed in his document, would win the "sympathy and active cooperation" of his readers. If it did, he noted, they would soon have the opportunity to sign a petition of support and attend an organizational meeting on the matter. To Rogers's pleasant surprise, over two hundred replies pledged support for his program.[29]

Following the distribution of the *Objects and Plan* and circular, Rogers called interested citizens to a meeting in January 1861. Two important decisions emerged from the gathering. First of all, many of the supporters who had replied to Rogers's mailing attended the meeting. By popular vote they decided to seek incorporation and Back Bay lands from the legislature as soon as "legally empowered and properly prepared to carry these objects into effect." Second, members voted to establish Rogers as chairman (and twenty-first member) of a Committee of Twenty "to secure a grant of land" and "to frame a constitution and by-laws for the government of said Institute." Within days of the meeting Rogers had submitted to the legislature a list of supporters and a memorial for MIT.[30]

The *Objects and Plan* and the January meeting galvanized support from various quarters. Industrialists appreciated the potential supply of skilled laborers that the institute would produce. "The project has my warmest sympathy. . . . [for] there has long been a want," wrote William P. Blake, a Yale graduate and mining geologist, to Rogers, "of young men fitted to take charge of ore mining establishments. . . . I hope to hear of the progress of this undertaking." Others with an eye on industry, such as James E. Olivier, saw the institute as a means to "educate our artisans up to the point of independence so that they can prosper alone if that be the prize of our determination to protect the personal rights of every citizen or stranger upon our soil." Some educational leaders concluded that MIT might address an important, unfulfilled need. President James Ritchie of the Mechanics Institute in Roxbury, Massachusetts, told Rogers of people "on all sides among every grade . . . who most anxiously crave opportunities like those which would be afforded by such an institution as you propose." Scholars, too, joined in support of MIT. Benjamin Peirce, Harvard's Perkins Professor of Mathematics and Astronomy, spoke enthusiastically of the plan. An attendee at the January meeting recorded that Peirce "addressed the meeting, heartily approving of the plan proposed, and regarding it as a much-needed institution. . . . He thought there was a great want of such practical education among our mechanics, and there was yet no Institute which could supply the want." Peirce spoke from experience, for he knew the difficulties of promoting such a system of practical education after witnessing the failure of his own reforms at Harvard. He attended the gathering in Boston to promote the idea of MIT, despite being on the faculty of the Lawrence Scientific School, a potential rival to the institute.[31]

Between the January meeting and the March session of the 1861 State Assembly, Rogers received additional support and met the expected opposition. From his earlier political lessons he had learned that he needed to build alliances for his plan as well as contend with the school fund. As his MIT proposal navigated the uncertain waters of the legislature, he happily attached several major "Petitions in Aid" to the project. One came from the Boston Society of Natural History, an institution that hoped to benefit from the plan for MIT. Such aid, the society declared, would help extend and perpetuate "its usefulness" to the state. The Boston Board of Trade, the American Academy of Arts and Sciences, the Massachusetts Charitable Mechanics' Association, and the New England Society also contributed words of support. Rounding out the petitions, State Teachers' Association members expressed their "hearty interest [in] and sympathy" with Rogers's proposal. The teachers' association lauded the MIT plan of education as being "suited to the development of intelligent industry and the promotion of liberal culture in connection with industrial pursuits."[32]

Buoyed by a cache of Petitions in Aid, Rogers then turned to the potential opposition from school fund advocates. By late March he worried that "the crisis of our Technological plan is now approaching." Fearing a repeat of the year before, he worked on stemming opposition in the Senate. "I am very busy," noted Rogers, "corresponding with persons of influence in different parts of the state, in order to give the Senators a true appreciation of our plans." As expected, resistance came from allies of the secretary of the Board of Education. Fortunately for Rogers, however, the former and powerful secretary, George Boutwell, had been replaced by a "newly appointed officer of little influence." Although the new secretary opposed Rogers's plan, "the Board itself has not sustained his opposition. Some of the Professors at Harvard have shown sympathy with us, among them Peirce, Bowen, Judge Parker, etc."[33]

Presentations to the legislature on behalf of the MIT plan helped secure the sympathy of several state leaders. Rogers counted on one presentation in particular to defuse the concerns of school fund advocates. A report prepared and presented by M. D. Ross provided an *Estimate of the Financial Effect of the Proposed Reservation of Back Bay Lands.* Rogers and his circle had commissioned the report to evaluate alleged losses to the school fund that would be caused by issuing a land grant to the MIT proposal. Through the report Rogers argued that, "by making the reservation, and granting use of it to the proposed collection of institutions, the adjacent lands will be doubled in value." He made use of a detailed sketch of recent land acquisitions, developments, and subsequent land value increases of the Back Bay to make his case. The presentation, among others, helped overcome resistance from some school fund supporters.[34]

During the closing deliberations of the legislature over the MIT proposal, Governor John A. Andrew warned Rogers that lingering obstacles still threatened the plan. Although several able speakers had, over the course of the session, come to the defense of the proposal, Andrew, a friend to Rogers, believed only one voice could prevent a third defeat. "I hope you will come and advocate the claims of the . . . Institute of Technology," he wrote to Rogers, "but no one else should speak. Be thou the advocate." He not only followed the governor's suggestion but also wrote the Joint Standing Committee report for legislators assigned to review the MIT proposal. The report restated the requests and assertions Rogers had made in the *Objects and Plan.* In addition, it contained two stipulations. First, the legislature was to grant both a charter for the institute and, if proponents of the plan succeeded in raising a $100,000 endowment, plots of land on the Back Bay. Rogers had no qualms at the time with compromising on the fund-raising stipulation. Earlier he had even asserted in private that "within a year, two or three hundred thousand dollars will be devoted to these practical objects should the State make the grant for which we ask." The second stipulation that emerged in the final bill, however, went over less smoothly. This second requirement, which Rogers called the "ungracious condition," had to do with the school fund. Advocates of the fund had managed to insert a condition that, if land values surrounding MIT did not increase as projected, Rogers and his circle would be required to repair the loss.[35]

After following the governor's instructions and drafting the committee report, Rogers felt assured of a victory shortly before state leaders voted on the institute's fate in late March 1861. "In the Senate," he wrote to Andrew, "I think there will be no serious obstructions—several who opposed last year having said they would help us now." With those words of confidence, Rogers accurately predicted the actions of the Senate that sent his long-held plan for higher educational reform to the governor's desk for signing.[36]

Almost with a sigh of relief, Governor Andrew signed the Act of Incorporation for the Massachusetts Institute of Technology on April 10, 1861. The act granted Rogers and his circle full legal authority to incorporate and proceed to organize according to the *Objects and Plan*. Attached to it were the two seemingly minor stipulations that observers dismissed at the time. If they thought their difficulties were now behind them, they were sadly mistaken, for in the days that lay ahead the world as they knew it was about to change.

CHAPTER SEVEN

Convergence of Interests

ROGERS HAD ONLY two days to celebrate the start of the Massachusetts Institute of Technology. On the morning of April 12, 1861, an old friend of his helped bring about an abrupt end to any optimism then occurring in the nation. That day Edmund Ruffin, Rogers's old "marl" partner from his Virginia years, had woken up in a tent on the shores of South Carolina. Although he was no longer the young man of his earlier days with Rogers, Ruffin was surrounded by young men from all over the state. His primary affiliation that morning was with the so-called Palmetto Guard, but everyone there knew him as the fire-eating spokesman for the South. They believed his rhetoric of secession from the North, on the need to form a new southern nation based on southern values. They shared his vision of an independent confederacy of states that would protect above all the sanctity of the institution of slavery. They cheered him on from his tent as he approached a canon and fired a shot out into the water toward Fort Sumter. Had it been any other time or place, the blast wouldn't have meant much. But as it turned out, he'd fired the first shot of the Civil War.[1]

For Ruffin that day marked the start of a glorious campaign toward independence. For Rogers it meant that MIT would remain a mere idea for several years to come. During the Civil War Rogers nevertheless continued to be active in science, professionalization, and the promotion of the Institute. These activities gave him reasons to remain optimistic during and after the war. Scientists of the useful arts persuasion received unprecedented attention from the public as a result of the conflict. "Innovation is now very active in the war direction," Rogers remarked, "and everyday discloses some new scheme for defence or destruction." By the end of the war the useful arts had made an indelible impact on the public mind. Science, as applied toward military ends, had played a vital role in determining the outcome of the conflict. Industrial power, as associated with scientific and technological innovation, was perceived to have shaped the course of human events.[2]

But the war upset the balance of much of what Rogers and the nation had taken for granted. Even afterward, he found difficulties advancing the Institute's cause.

Whether raising an endowment or translating his idea into practice, Rogers's path toward establishing MIT met no shortage of challenges. While facing these hurdles during and after the Civil War, he remained a staunch advocate of the useful arts ideal. As the scientific, professional, and educational interests Rogers held began to converge, he steadily sought to balance practical and theoretical interests.

A mere two months after the fall of Fort Sumter marking the end of the war, the state of Massachusetts asked Rogers to apply his scientific experience for practical ends. The legislature passed a bill authorizing Governor John A. Andrew to appoint an inspector of natural gas usage in the state. At first Rogers declined the generous offer, that included a comparatively large annual salary of three thousand dollars, an office, budget for experiments, and assistants to monitor the gas infrastructure. He had his mind on other matters, such as developing MIT and meeting the endowment requirements of the charter. The state appointment, he feared, would distract him from his work on the Institute. He offered the governor the names of two other scientists who could fulfill the duties of the position, John R. Rollins of Lawrence and A. A. Haynes. As specialists in chemistry, they would, he suggested, match "the business capacity, integrity, education, and scientific taste" needed for the position, and his "own inexperience with business matters" made him less qualified for the post. Andrew, however, refused to accept Rogers's decision on the matter. The governor held a meeting with him and Haynes that ultimately persuaded Rogers to reconsider the offer. Rogers left convinced that he had everything to gain and little to lose by accepting the position. Acknowledging Rogers's concern about being drawn away from the Institute, the governor assured him that at most the office would require only one or two hours per day of supervision. Assistants, Andrew made clear, would absorb most of the work.[3]

Rogers accepted the assignment and was quickly satisfied about his decision. He would have an office, he reflected, and "at the expense of the state, a perfect and complete set of standard gasometers, photometers and other essential apparatus of the most perfect patterns." He believed that being exposed to the best equipment then available would help him decide on the best equipment to get for MIT. He also envisioned having an "expert assistant" or two, with whom he could conduct research "sufficiently scientific not to be distasteful." The new state office intersected with Rogers's ideals about the function of science for social ends. As he put it, the position would "stand between the consumer and gas companies." If his task was to ensure public safety and advocate a rational approach for the distribution of a natural resource, he was sure to find professional satisfaction, not to mention valuable professional experience in the new post. In the end he expected "this office not to interfere much with [my] general science, and it may help on my Technological plans." He decided to give it a year to see how it went.[4]

Between accepting the new position in late June and the second day of September 1861, when he began sending assistants into the field, Rogers looked for ideas abroad and at home to prepare for the task. In much the same way he corresponded about professional or educational matters, he turned to his brother Henry in Europe to learn about gas inspection. He began by asking Henry for literature related to illuminating gas. Wanting to be "armed in the completest manner possible," William requested papers, documents, and advice on recent apparatus developed in Europe. Having trouble finding a well-known work, the *Gas Inspector's Manual* he asked Henry to locate a copy in Great Britain. William also turned to domestic examples from which to learn about his new responsibilities, particularly in New York. There he expected to visit the "Gas Works laboratory," which conducted similar work to what he planned to do in Boston. He wanted to take a few days "to look at the admirable arrangements . . . under [John] Torrey's charge."[5]

From documents sent by his brother and from experiences in New York, Rogers gained enough confidence to start a career in gas inspection. For nearly three years he set about verifying standards and delegating "meter-proving" to his assistants. He also assembled an arsenal of measuring equipment imported from London for estimating precisely the amount of gases distributed, released, or otherwise in use across the state. Although he lamented at times that his "gas engagement precludes much attention to purely scientific matters," he took a serious interest in the operation and technologies of his office. From the post he began to urge fundamental changes to gas inspection that aimed for greater accuracy in measurement and improved public safety. One of the changes he recommended was examining gas meters by way of air, rather than gas, for controlling air was easier than controlling gases. By examining the meters with air, more precise data could be analyzed.[6]

Rogers accepted the role of inspector, hoping it would not consume much of his time. But within weeks of starting his duties, he became mired in the politics of his job. Most of the political turbulence he experienced involved two types of meters, wet and dry. Many small companies in the state had invested in wet meters. Gas companies and meter makers, meanwhile, preferred dry meters and had combined their political influence to exclude wet meters from use. Rogers found himself caught between these competing forces, for he had the authority to register or exclude whichever system he chose based on evidence he examined. "By the refusal of my seal and stamp," he remarked, he realized he could damage the interests of one group or another. To help him with the decision, Rogers requested documents from Europe on the "merits or demerits" of wet meters and proceeded to have his assistants test approximately two thousand dry meters. While a majority of the dry meters passed Rogers's test, he failed to pass many others. With his new instruments of measurement, he found that the previous methods used by the state were "very imperfect and fallacious." Rogers's scientific observations, however, had begun to jostle a political beehive. "A New York man, one of the largest manufacturers in the country," Rogers commented in December 1861, "is now here for the purpose of learning why so many of his meters have been rejected by me, although sealed and stamped by the New York inspector." Rogers planned to show the manufacturer tests that indicated flaws in the meter, hoping the tests would prompt improvements in the construction of meters. Following this incident with New York, Governor Andrew requested a report from Rogers. The report satisfied the governor and Rogers continued to be absorbed by the duties of his office. He later noted with pleasant surprise that "the leading manufacturers [there] are adopting my methods and standards." With a record of improvements made in his first year, Rogers planned to resign the position and concentrate on MIT.[7]

The end of the first year came and went, however, without a resignation. Instead, Rogers was further drawn into the useful arts character of his work. In keeping with previous interests in technology development, he began to devise instruments to help him with his inspections. With assistance from local instrument makers, Rogers improved on test meters by "little contrivances" of his own construction: "I am just now completing a portable photometer . . . for comparative observation. The contrivances hitherto used for the purpose are very unreliable, but this will, I think, prove satisfactory." He began to ask small companies to adopt the use of his instrument and noted that large companies would be required by law to have them. Toward the end of the second year as gas inspector, he proposed some research studies on gas: "I have been making some interesting experiments in my office upon the effect of different quantities of carbonic acid contained in coal gas upon its illuminating power." Publications on the topic followed, but his time as inspector ended on February 1, 1864, when Rogers resigned. He told the governor of his plan to "devote myself to those educational plans which are hereafter to make large demands on my energy and time."[8]

After this period in the inspector's office, the state tapped Rogers for another science-related project. This time the duties were of a brief duration and involved an opportunity to travel abroad. Governor A. H. Bullock appointed Rogers to represent Massachusetts in the Universal Exposition of Paris in 1867. International exhibitions of the nineteenth century first began in midcentury London out of the desire to display the fruits of European industrialization. These fairs produced grand-scale exhibits of innovations in communication, transportation, and industry. They also sparked competition between nations in these perceived markers of progress. Situated on forty-one acres along the Seine, the 1867 exposition was organized around the theme "The History of Labour." Rogers prepared for his visit to Paris, expecting to witness the development of machines of industry from the Stone Age to the latest manifestations of industrial society.[9]

Accompanied by his wife, Emma, and MIT chemistry professors Charles Eliot and Francis Storer, Rogers left for France in June 1867. Along the way, he planned excursions to see manufacturing innovations in Great Britain, such as "a new furnace, of which an account was given sometime since at our Institute, and which I was very desirous to see in action." The originals of Watt's steam engines, "which are still in daily use," also captured his attention. Author Gustave Flaubert, also attending the fair, remarked: "It is overwhelming. It contains splendid and exceptionally curious things. . . . Someone who had three whole months at his disposal to visit the Expositions every morning and take notes could spare himself the trouble of ever having to read or travel again." Rogers had precisely this kind of time at his disposal, and the work there consumed him, leaving little for anything but notetaking on the collections and presentations. With assistants on hand, Rogers set out to study "some of the departments of the useful arts, as here illustrated, with a view of gathering material for a Report." The problem his team encountered had to do with the gargantuan size and scope of the exposition. "I have but one fault to find with it," he commented, "which to superficial observers is, I suppose, its highest merit,—it is too vast." An enormous basilica housed the main exhibits, arranged in such a way that observers could study innovations by industry in the "galleries" or by nation along the "avenues." Outside stood a seemingly endless stream of smaller displays. He wished, instead, that only the "really new or original" was presented there, as he sifted through the collections of machines of production, science, and industry. The overwhelming number of exhibits and the sea of visitors, ranging from forty to sixty thousand individuals per day, made the task of gathering information for Massachusetts more difficult than he had expected. Still, Rogers concluded that the exposition "far transcends in richness and extent all that I had imagined."[10]

From the visit to the Paris Expo, as from his work as state inspector, Rogers acquired valuable experience. These duties, the war, and emerging concerns about his health, however, limited his opportunities for doing scientific research. He wrote occasionally on such topics as new technologies and processes for measuring illuminating gas or descriptions of the measurement of electrical illumination in Boston. But during and shortly after the Civil War he shelved his larger program of research in geology and natural philosophy. Inspection work, the national conflict, and the trials of aging became increasingly burdensome, taking a toll on his health. Seldom specific about the ailments that plagued him, he stated, "I expect for some time yet to pay the penalty of my forgetfulness of this constitutional peculiarity." His decline seemed particularly worse during periods in which his appointments demanded much of him and limited the range of interests he could follow. Thus, his program of research gave way to other activities in his profession.[11]

While the number of his research efforts fell, Rogers's involvement with organizing efforts continued. Not all of the appointments he received, however, were expected. In the antebellum years Rogers's active participation in the founding of the American Association for the Advancement of Science made service appointments to those institutions understandable. Yet from organizations founded after the start of the war, he received two elected positions that took him by surprise.

One of the positions offered came from the American Social Science Association (ASSA), established in 1865. During its founding and formative years the ASSA had emphasized on investigatory commissions and civil service reform. With this intent in mind, members of the association voted for Rogers as their first president. Organizing members had at least four probable reasons for selecting him. First, Rogers symbolized the broad effort of professionalization in the United States. As a participant in the formation of the Association of American Geologists and Naturalists and, later, the American Association for the Advancement of Science, he had experience that they believed might serve them well. Second, as a published natural scientist, he represented a form of scientific inquiry that social scientists sought to adopt for themselves. Having conducted the Geological Survey of Virginia, published in the field of mechanical philosophy, and engaged in debates over evolution, Rogers's work followed recognized patterns of scientific research. His internationalism gave members a third point of common interest. Founders of the ASSA depended to a large degree on models for similar organizations in Britain. Rogers's association with British scientists and affiliation with several of their scientific societies would have had an inherent appeal to ASSA members. Perhaps more important than anything else was his interest in the useful arts. That he believed in a form of science that could be applied for practical purposes—while also retaining the credibility and authority of traditional science—likely attracted members of the social science community. In this light, Rogers, as their first president, would have been perceived as being helpful in establishing a scientific approach to social reform efforts. His previous activities as an organizer and promoter of science, defender of Darwinism, and advocate of the useful arts thus enhanced his appeal as a candidate among social scientists searching for leadership.[12]

In the end Rogers overcommitted himself and barely participated in ASSA activities. Most of the duties fell to such officers as recording secretary Frank Sanborn and corresponding secretary Samuel Eliot. State and MIT duties led Rogers to submit a letter of resignation in 1866. Sanborn, among others, persuaded Rogers to stay on for a time to provide continuity for the association.[13]

If the ASSA position took Rogers by surprise, another with the National Academy of Sciences (NAS) came as more of a shock. The academy began as an idea in the minds of select members of the Lazzaroni. Louis Agassiz and Alexander Dallas Bache, in particular, spearheaded the secretive launching of the organization. Working closely with Senator Henry Wilson of Massachusetts, Agassiz and Bache had shepherded a bill through Congress to incorporate the organization in 1863. The Civil War era called for some organizing body, the bill argued, to give oversight to the developments of science and technology research. To bolster the organization's power and prestige, they claimed that only the most recognized scientists should be affiliated with the academy. The primary function, at its start, would be to give advice to government leaders on how best to spend its resources for the advancement of science and technology. Within a few years of its founding, the academy was fulfilling its role. The secretary of the Treasury requested recommendations on a uniform system of weights and measures, on protection from counterfeit currency, and on alternative metals for coinage. The navy requested solutions to problems associated with saltwater and its iron ships. Other government requests included critical reviews and evaluations of research as well as instruments.[14]

The secrecy surrounding the NAS's founding, however, left an uneasy impression on Rogers and many other scientists. To the surprise of even some of the Lazzaroni, scientists such as Joseph Henry, who claimed no knowledge of the effort, the bill passed in a late-night session of Congress and later was signed into law by President Lincoln in the spring of 1863. Of the fifty incorporators, chosen by the Lazzaroni, Rogers believed several had, beyond their friendship with the founders, dubious qualifications; he also lamented that other scientists of prominence were excluded. In his view the NAS, while alleging objectivity in the selection of its members, marred its own launching by its favoritism.[15]

Rogers appeared among the list of incorporators and must have found it puzzling, considering his disagreements with the Lazzaroni over previous professionalization efforts. He questioned the process of incorporation, in which "only two or three of the men of science knew anything until the action of Congress was announced in the papers." He wondered also about the composition of the list of incorporators: "[Josiah] Cooke and [Joseph] Lovering are left out, though many an unknown name is placed on the roll of honor." Nevertheless, he went to the inaugural meeting held in New York in April 1863 and, from the start, was swept into the politics of the gathering that met at New York University. On his way to the meeting room, he met briefly with John William Draper, professor of chemistry at New York University, who hadn't received an invitation. Draper, by this time, had a well-established reputation as a scientist who, like Rogers, was interested in both practice and theory. His published works included textbooks on chemistry and natural philosophy; his work in the area of photochemistry yielded gains in the field that made early forms of photography possible. He also became known for the Grotthus-Draper Law, a theory about the relationships between the absorption of light and chemical change. "Surprised" and "mortified" at the fact that NAS policies prevented Draper from appearing on the incorporator's list, Rogers nevertheless proceeded to the meeting. According to him, members opened the first session by indulging in "exultation and mutual glorification." Dissatisfied with this behavior and unafraid to speak his mind, Rogers began to raise concerns that he believed others had as well: "This . . . is a sad error," he told the attendees, "if it be not a grievous wrong. Surely . . . there are many here who in their hearts must feel that they have no claim to be here when such men as I have named have been excluded." The names he referred to included not only Draper but also astronomer George P. Bond, celebrated by the Royal Astronomical Society of London, Elias Loomis, on the science faculty at Yale, and Spencer Baird, zoologist and administrator at the Smithsonian Institution.[16]

Aware that friends Asa Gray and Theodore Lyman had chosen not to attend the meeting, Rogers began to question whether he should've bothered to be there at all. That only twenty-seven of the fifty incorporators stayed for the entire session in New York to organize the NAS added to his doubts. He soon discovered, however, that the burden of resisting the Lazzaroni fell upon him. They brought the "most objectionable provisions" to a vote, and he, virtually alone, stood in their way. Rogers, serving on a nine-member committee of organization led by Bache, waited till the very last moments of the opening session to deal with the issue of terms for NAS officers. Astonished that no one opposed life tenure for the offices of president, vice president, and secretary—Bache, James D. Dana, and Joseph Henry, respectively—he "let it pass without voting until, the morning's task being closed, Bache was about shutting up his book. Then I rose, and calmly called their attention to this clause, told them that to exact that would be to blast every hope of success, and so impressed them with the responsibility of such a course that they voted the term of six years instead of life." Although opposing the Lazzaroni had required "much use of [his] backbone," Rogers found support among those gathered there, particularly from J. S. Newberry, director of the Ohio Geological Survey, and Stephen Alexander, astronomy professor at Princeton.[17]

The Lazzaroni received a bitter reminder of why they had broken from the American Association for the Advancement of Science in the first place. Rogers had pushed for democratic reforms, leading an anti-Lazzaroni movement that desired to expand its power in decision making and the election of officials. Now his presence promised similar tensions over the elite-run NAS. Benjamin Peirce of the Bache clique feared a backlash against the NAS. Following the meeting in April, Boston's American Academy of Arts and Sciences voted to elect officers. "To show their hatred of the National Academy," commented Peirce to Bache, "all its opponents combined to elect Gray as President and William B. Rogers as Recording Secretary" of the Academy of Arts and Sciences.[18]

In reversing Bache's attempt at a tenure-for-life presidency of the NAS, Rogers touched off internal conflict. It didn't take long for him to assume leadership of an opposition group. "I want to talk Academy to you," mentioned Rogers supporter J. S. Newberry. "As you will have learned, it will be expanded and rendered more democratic and popular at the next meeting or expire. Which shall it be?" Newberry referred to the passing of leadership from Bache to Joseph Henry in 1865. Bache had become seriously ill, leaving executive duties to Henry, who reluctantly assumed the role after Dana resigned the vice presidency. With the Lazzaroni's hold on the NAS eroding, Rogers's circle became more assertive. Louis Agassiz soon found his attempt to reject the membership of zoologist Spencer Baird opposed by a majority of naturalists threatening to resign from the academy. By the end of his career Baird had over one thousand publications to his name, including major works on reptiles, mammals, and, his specialty, ornithology. Agassiz's fight against Baird raised more suspicions about personality conflicts than it did questions of Baird's credentials. Henry felt he needed to remind Agassiz that in "this Democratic country we must do what we can, when we cannot do what we would. We must expect to be thwarted in many of our plans." Over the years following Henry's assumption of NAS leadership, Rogers's cohort supported a shift in policy that made "original research" the basis for membership.[19]

＊＊

While Rogers tinkered with state science projects and professionalization activities during and after the war, he spent the bulk of his time attempting to keep the idea of MIT from evaporating. The war, however, made fund raising for education a difficult task. The same week bullets began to fly, the state gave him a one-year deadline to raise $100,000, a substantial sum for even flush times. As the conflict wore on, his prospects only worsened. He turned to two potential sources of stability during the first half of the war, one local and the other national. In the process he found that the philosophical and philanthropic underpinnings of the

MIT idea clashed with those of Louis Agassiz and the Museum of Comparative Zoology.

Rogers first turned to private sources of support among Boston area philanthropists. Over two hundred citizens from Boston, Cambridge, and neighboring cities had registered their interest in MIT when the proposal went before the state legislature. Many of them donated small sums to the project during the first year of the war. A few offered large gifts. Among the most generous was Ralph Huntington's $50,000 allotment to MIT in his will. Huntington, an industrialist, president of the Boston and Roxbury Company, and supporter of the idea of MIT, had been approached by Harvard for support, but he preferred Rogers's plan. Even so, when the deadline of April 10, 1862, arrived, Rogers had raised a mere fraction of the endowment requirement. Promises of future income, such as Huntington's, may have brightened Rogers's prospects, but the legislature wouldn't recognize funding of this sort for the deadline at hand. Thus, two days before the legislature was scheduled to revoke the charter and take back the lots, Rogers convened an emergency meeting of MIT organizers. At the meeting he proposed drafting a formal statement accepting the charter and land while also petitioning state leaders for a one-year extension to raise funds for the endowment. Organizers agreed to the move. Rogers sent the acceptance and request and then waited anxiously to hear the state's reply. Given the economic uncertainties produced by the war, the state assembly decided to grant Rogers another year.[20]

Relieved as he was at the extension, he still wondered whether another twelve months would make any difference. The first year of the war had come and gone; young Bostonians, who might have enrolled at Harvard or MIT or elsewhere, prepared for military service rather than for their studies. Whatever money then circulating was being directed toward defeating the Rebels rather than at starting an institute.

Following the extension, Rogers organized MIT's first official meeting on May 6, 1862. Institute members met to establish a constitution, bylaws, and a "government," or "corporation." The governing body was composed of a mix of intellectuals and industrialists who agreed to develop two of the three parts outlined in the *Objects and Plan*. The Society of Arts and the School of Industrial Science would come first; the museum, they decided, would come later. Also at the meeting, the government elected the Institute's officers, including Rogers as president and a handful of vice presidents. Appointed to organize and oversee the first few month's of MIT's activities, Rogers moved forward with the Society of Arts, the research branch of the Institute. By December he gave the introductory address that launched the bimonthly conference on developments in the useful arts. The society engaged all manner of theoretical and practical issues related to "the mill, the farm, the machine shop, the laboratory, the shipyard, from the desk of the engineer and architect, the chair of science, the workman's bench, the merchant's counting room and all the other scenes where educated industry is at work." The papers that followed in this and other meetings of the society dealt with a wide range of interests. Some reflected the military problems of the Civil War era, with presentations on "Sub-aqueous Gun Firing." Others applied to perennial problems in the useful arts. The Society of Arts lasted eight years, and Rogers all the while kept active in this branch by preparing over thirty research-based presentations for its members.[21]

However much Rogers may have enjoyed participating in Society of Arts meetings, he still fretted about the looming deadline. He had raised virtually no new funds and began to worry whether he could meet the charter's requirements. With local purse strings held tight, Rogers considered other avenues of support. If he followed the developments in Congress at the time, he would have learned that during the summer of 1862 the federal government had passed the land-grant appropriation act. The bill was hardly a new idea. Vermont senator Justin Morrill had talked for years about the need to establish federal support for agriculture and mechanics education. Because such proposals often failed at the local level, Rogers might not have thought much of the bill at the national level. When Morrill successfully pushed the bill through Congress and received the president's signature in July 1862, Rogers became more curious. He began to wonder whether MIT could apply for land-grant funds. As mandated by the bill, states would receive thirty thousand acres of land in western territories for each of their congressional representatives. Consequently, the formula favored the more populous states, which received larger appropriations. Morrill's bill allowed for states to then sell, rent, or otherwise derive an income from the scrip to finance the establishment of a new institution or to support existing ones. The legislation, however, required that those institutions receiving the funds must promote the education of agriculturists and mechanics of the state. Rogers had little reason to doubt that MIT could fulfill such a role. Indeed, one of the guiding premises of the Institute was the study and advancement of the mechanic arts.[22]

In December 1862, the same month the Society of Arts held its first meeting, Rogers's counterpart in science and professionalism, Louis Agassiz, began to maneuver for land-grant funds. Agassiz began to correspond with Governor John A. Andrew about how Massachusetts could best use the appropriation. At first Agassiz complained in vague terms that Harvard had "more the character of a high school than a University." The Lawrence Scientific School, he asserted, had the potential for greatness on par with the state-supported universities of Europe. As it was, however, detractors could legitimately call the Science School mere "excrescences of the college proper." In short, Agassiz argued that state support could raise the level of science at Harvard to match that of European institutions. He followed his complaints with more specific appeals for funds to support agriculture education, of which Harvard had little previous experience. Agassiz, by this time, had heard of a legacy left to his institution for the founding of the Bussey Institute of Agriculture. He used this development in his plea for the federal grant. "You might make a good beginning," he remarked, "toward founding a University by combining your resources for the organization of an Agricultural College with those of Harvard to which a large legacy has been left for a similar purpose." Coupling the $200,000 Bussey bequest with the Morrill funds, he assured the governor, would have far better effect than dispersing the funds across several institutions. By this Agassiz was referring to MIT and other potential candidates that might compete for the grant. He warned Andrew against dividing the funds with others, an act that "will provide nothing above mediocrity." Harvard would comply with the Morrill legislation, he maintained, even with military drilling as required by the bill. Indeed, Agassiz went as far as to ask the governor to request such a program at Harvard ahead of any decision on the distribution of funds.[23]

Of all the hinting and prodding, Agassiz's next proposal struck Andrew as the most appealing. Drawing on the governor's interest in MIT and hoping that Andrew might want to make a name for himself in American higher education, Agassiz began to suggest a merger between MIT and Harvard. If Agassiz could not convince the governor of Harvard's commitment to mechanics and agriculturists, perhaps, he suggested, Harvard should simply acquire MIT for the purpose. I understand, Agassiz mentioned, "that the gentlemen who have contemplated the organization of a polytechnic school propose to press their scheme this winter." If this is the case, he continued, "the opportunity should not be allowed to pass without making an effort to combine this plan with whatever may be done for an agricultural college and towards the founding of a great university." When Agassiz penned these words, a mere five days had passed since the Society of Arts held its first meeting. MIT now posed a threat to his plans to draw all the land-grant funds for Harvard.[24]

For every ounce of desire Agassiz might have had for the funds, Rogers could have matched five ounces of conviction that MIT should remain independent. If Rogers had known the Agassiz scheme, he would never have entertained it for long. Agassiz was likely aware of this, which explains why he went to Andrew about a merger rather than to Rogers himself. Rogers learned soon enough, however, about scheming of one sort or another from Andrew. In late December the governor asked Rogers to prepare a report on MIT's progress. Andrew wanted to discuss some of the developments with the Institute at his annual address in January 1863. In the same request the governor also asked his opinion on a plan for the future of MIT. Without revealing the origins of the plan, Andrew asked Rogers whether combining Harvard and MIT might bring about good effect when coupled with the Morrill appropriation. As requested, Rogers replied with a detailed progress report on events related to the Society of Arts and the organization of MIT's administration and governance. As for his opinion on the merger proposal, Rogers stopped the governor's inquiry dead in its tracks. Without hesitation he replied to Andrew that "the institute had from the beginning determined to stand alone, that its independence was essential to its success, and that it would accept no grant . . . which should in the slightest particular interfere with this independence."[25]

Rogers brandished a bold response, considering he faced an empty coffer, a looming deadline, the prospects of losing MIT's charter and lands, and no real alternatives for meeting the endowment requirements. But central to the idea of MIT was the notion of autonomy, a freedom to experiment with instructional methods, the curriculum, and approaches to science. Merging with Harvard, to Rogers's mind, would bring the end of such autonomy, along with rigid traditions and, ultimately, the likely absorption of the Institute into the patterns of the Lawrence Scientific School. Not far removed from these concerns were the fundamental differences he had with Agassiz. Rogers had already seen how Agassiz had succeeded in transforming, at least in part, the original, largely applied mission of Lawrence's program into an extension of his research that had little to do with utility or application. Agassiz lacked sincere interest in the useful arts and considered polytechnic schools an intermediate between trade and science, somewhat like "high schools, which are the necessary medium between the primary school and the university." They had also been arch rivals in professionalism and science. The two had sparred with each other over the parameters of professionalization in science organizing. Where Rogers wanted open, democratic access to the election of officers, Agassiz valued a seemingly oligarchical approach. As for scientific research, Rogers favored the useful arts, the laboratory method, and Darwin's views as expressed in *Origin of Species*. Agassiz emphasized natural history through museum work and the collection of zoological specimens, and he considered Darwin's speculations anathema.[26]

When it came to science in higher learning, both men valued observation as a mode of instruction, although through entirely different means. Agassiz believed in a view of science Rogers would have called "statistical," one that emphasized counting similarities and differences between the structures of organisms for the purposes of cataloguing and classification. This approach required extensive museum collections and, as in Agassiz's case, relied to a great extent on the naked eye for observation. "Agassiz used to lock a student up," recalled psychologist William James, "in a room full of turtle shells or lobster shells or oyster shells, without a book or word to help him, and not let him out till he had discovered all the truths which the object contained." According to another student, Samuel H. Scudder, "instruments of all kinds were interdicted. My two hands, my two eyes, and the fish: it seemed a most limited field." Agassiz expected students to develop the faculty of observation by spending days examining the exact external features of a particular specimen. His philosophy of education centered on teacher silence. To awaken the faculty of observation, he argued, "I must teach and yet give no information. I must, in short, to all intents and purposes, be ignorant like you." Agassiz, in large measure, hoped to prepare his students for the vast cataloguing work that stood before him at the Museum of Comparative Zoology. While it would be misleading to assume that no microscopes or study of internal structures, such as embryology, were used in the museum's taxonomic work or Agassiz's research, the primary instructional mode was not experimentalist in nature and, instead, favored observations that kept structure and composition intact.[27]

Rogers, of course, had long advocated the use of laboratory instruction. He envisioned the decline of natural history and its "statistical" goals along with the rise of natural philosophy and the experiment. Subjecting structures of all kinds, organic and inorganic, to experiments, observing their reactions, and considering the relationships between the structures and those reactions held the keys to the future. Instruments, he believed, offered limitless opportunities to explore the parameters of nature, to discover new methods of instrumentation to probe the natural world, and to build new taxonomies for understanding natural phenomena. He was also convinced that these developments would bring about practical applications that should be investigated. MIT stood for each of these values through its commitment to the laboratory method. To Agassiz experimentation of this kind had limited value for the advancement of science as he understood it. For him a true taxonomic system, based on untampered structural characteristics, had already been established, and newer ideas offered distractions from the natural historical goal of science.[28]

Harvard president Thomas Hill, a friend of Agassiz, also competed with Rogers when he wrote to Governor Andrew about the Morrill appropriation. Days before the 1863 annual address, Hill discussed with Andrew the ways in which federal funds would benefit the Science School and the Bussey Institute, mentioning nothing of the merger proposal. His silence, perhaps nothing more than an oversight, may have reflected an uneasiness with Rogers's useful arts approach to science. Rogers, whose scientific research had involved traversing rugged mountain terrain, enduring adverse weather, and soiling his hands with earthen materials and laboratory chemicals on such projects as the Virginia survey, was likely to encourage his pupils to do the same. Hill, meanwhile, warned students against overexerting themselves in physical labor, claiming to have injured a testicle while weeding his garden. Rogers's association with Darwinism would not have helped Agassiz's merger plan either. Hill, a former Unitarian minister, held as much contempt for evolution as its staunchest critics. Whatever the case, Harvard's president didn't actively promote the merger scheme and became, as Rogers put it, one among many "rushing in to claim a slice of the loaf which comes to the State from the land grant."[29]

If Rogers's flat rejection of the plan and Hill's silence were intended to warn Andrew, the governor missed the message. He spoke at the 1863 address about the potential opportunities in combining MIT, Harvard, and the land-grant funds for the promotion of a true university. In short the governor sided with Agassiz, declaring a need to join "the Institute of Technology and the Zoological Museum." He imagined the two institutions "working in harmony" with Harvard College, securing "for the agricultural student for whom [Massachusetts] thus provides, not only the benefits of the national appropriation, but of the Bussey Institution and the means and instrumentalities of the Institute of Technology, as well as those accumulated at Cambridge." Agassiz might as well have delivered the address, for it followed the zoologist's plan to the letter.[30]

State leaders began mulling over the idea of a merger, which made Rogers feel increasingly uneasy. He never discounted "secret forces [that] continue to avert present action." By "secret forces" it is unlikely he meant anything other than Agassiz's circle and their plan that had gained the attention of Governor Andrew. Aware of the interest in a Harvard and MIT, Rogers went before the legislature to plead his case for the Institute's independence. In a meeting lasting several hours Rogers discussed with state officials the functions of the Institute and requested "one half, or at least one third of the proceeds." He understood the constraints of the bill, that none of the funds could be used for the construction of buildings, and knew that a separate MIT fund for the purpose had accrued ten thousand dollars.[31]

His unease continued, however, and for good reason. When a legislative committee issued a final report, its members still hadn't reached an agreement. The majority of the legislators praised the governor's plan to merge Harvard and MIT. The plan, they acknowledged, had been "warmly commended by many of the leading men of the state." A minority within the committee responded similarly at first, repeating the governor's statements against dividing the funds across institutions. But ultimately, the minority opinion rejected the governor's merger proposal, basing its decision on presentations made by agriculture and trade representatives. In the final report recommendations by the entire committee came in three parts. The state, first of all, should accept the land-grant appropriation from Congress. Second, the land scrip should be sold at market value, the proceeds of which should remain in a trust. And, third, the state should distribute the interest from the trust according to a predetermined formula. One-tenth of the interest, the committee recommended, should provide for the purchase of land on which to build an agricultural college. Of the remaining interest, one-third would fund MIT, and two-thirds would go to the development of the agricultural college. The report appeared in March 1863 and became law the following month. Rogers, while not interfering with the basic interests of the agricultural community, had managed to preserve the idea of an independent institute. For the agriculturists Amherst made more sense than Boston. The rural site and land-grant funds gave rise to the University of Massachusetts system.[32]

In order to accept the proceeds, Rogers's *Objects and Plan* needed a few adjustments. MIT, as originally conceived, had no program of "military tactics." As required by the Morrill bill, any institution receiving the federal funds would have to provide students with military instruction. The Institute also had to open its governing board to three state leaders: the governor, the chief justice of the supreme judicial court, and the secretary of the board of education. The legislation required ex officio membership in MIT's governing board for each of the public officials. The final adjustment required an annual report for the governor, a report that outlined developments at MIT in conformity with the land-grant act. Based on the reports or otherwise, the state reserved the right to cease its yearly appropriation to the Institute if MIT failed to meet the requirements as mandated by the legislation. By the middle of summer 1863 Rogers had submitted a formal acceptance of the grant and its conditions.[33]

Agassiz had little reason to mourn his political defeat to Rogers for very long. The Museum of Comparative Zoology at Harvard continued to receive generous support from benefactors drawn to his mode of science and education. Governor Andrew, however, felt a sense of failure well after his campaign for the MIT-Harvard merger. His annual address of 1865 lamented the failed plan for a grand university. He claimed to have been overruled by "the better judgment of the Legislature as to the views which I had the honor to present at length" about the merger idea in 1863. "Although I remain more fully convinced than ever," he reflected, "after two intervening years, of their substantial soundness, I have felt it to be my official duty cordially to co-operate. . . . My own idea of a college likely to be useful . . . is one perhaps not yet to be realized." The governor repeated his lamentations to agriculturalists the same year. In addition, he hinted at his inability to challenge the political forces behind the Joint Special Committee recommendation. Then as now, he remarked, "I do not think that the views which I entertain upon the subject of an agricultural college are those which, at this moment are quite popular in the Commonwealth among the farmers." He felt certain that his personal ambitions to create what he considered a true university, as defeated by the legislature, were at odds with the desires of the agricultural and trade communities. [34]

For as pleased as Rogers may have been about the committee's recommendations, he had still made little progress on the endowment. When news about the Morrill funds reached him, he had only a few weeks left before the April 10, 1863, deadline. With the extension about to expire, the Institute had raised less than half of the required $100,000. Rogers circulated a desperate plea to New England philanthropists for support. Small donations followed, and Rogers promptly replied with letters of gratitude. But a seemingly insurmountable sum needed to be raised when, on the very last day before the charter was to expire, William J. Walker, a Boston-area physician and philanthropist, pledged $60,000. On that day the idea of MIT cleared its final legal hurdle.[35]

With the charter, lands, and endowment in place, Rogers could begin thinking seriously about the first day of MIT classes. He still lacked a definite plan for the School of Industrial Science, not to mention a faculty, building, or student body. Thus, for the eighteen months that followed the summer of 1863, he focused on assembling the school's needs.

Rogers spent part of this period preparing the *Scope and Plan of the School of Industrial Science,* the foundation for the Institute's curriculum. The goal of the school was to offer "instruction in the leading principles of science" in relation to the useful arts. He included in the document a detailed description of a two-part plan of instruction, one for special students and the other for regular students. The program for special students involved the outreach activities alluded to earlier in the *Objects and Plan.* For these students, and for the general public, the Institute would offer "opportunities for instruction in the leading principles of science, as applied to the arts." Instruction for the "public at large," which included interested citizens of "both sexes," would occur in evening lectures to avoid interfering with workday schedules. The lectures aimed at "such useful knowledge as they [special students] can acquire without methodical study and in hours not occupied by business." Elementary courses suited for public lectures would cover such areas as mathematics, physics, chemistry, geology, botany, and zoology. Each of the courses would emphasize the "facts and scientific principles which are of leading importance in connection with the useful arts." For these students Rogers required no examinations or laboratory exercises but expected common decorum, or, as he put it, "conditions and restraints," during public lectures. Unlike the regular MIT program, the plan for special students allowed only for lectures.[36]

Regular students, on the other hand, faced daytime schedules, a battery of examinations, and a wide range of laboratory instruction. Although courses offered during the day occasionally resembled offerings provided in the evening, daytime classes employed a program of "systematic and professional instruction." Regular students would follow a partially prescribed, partially elective curriculum designed to prepare them for scientific and practical fields. Students could specialize in one of five program areas, including architecture, chemistry, geology, and two kinds of engineering (civil and topographical; mechanical). To this Rogers and the Institute would later add a sixth degree in general science and literature, for those wanting advanced theoretical training. In concert with the useful arts ideal, the plan aimed to offer students theory in the first two years followed by practical studies in the remaining two. To enroll in any program area, applicants no younger than sixteen would have to pass an entrance exam. Once enrolled, students faced monthly, midterm, and final examinations on material covered in the coursework. Rogers also required degree candidates to take comprehensive tests covering courses from all four years and prepare a thesis, although he allowed for some flexibility for experienced students in the regular program. Recent graduates of other colleges and universities, for instance, could apply for advanced standing, if they could show proficiency in the required introductory coursework. In some cases regular students could complete the prescribed course in three years. For the most part, however, those with college-level science experience would still find distinct academic requirements and challenges at the Institute. MIT differed from other science schools by emphasizing laboratory instruction and a comprehensive program of study that allowed for specialization. Rogers expected that his proposal would thus draw students or graduates from other institutions and assumed that most of them would enroll for the entire four years.[37]

He left little doubt about the centrality of the laboratory for regular students. "While attending lectures on the various branches" of science, he made clear, they "will have the benefits of laboratory exercises in manipulation and analysis." Rogers promised students a "practical familiarity" with the apparatus of the day through these exercises. The instruction would include the use and adjustment of laboratory equipment, experience with the materials commonly analyzed, and training in "the more important experiments and processes in natural philosophy and chemistry." Small classes would receive direct guidance from faculty in preparing, executing, and analyzing work in the experimental sciences.[38]

To this end Rogers described four specific laboratories that the Institute would establish: the Laboratory of Physics and Mechanics, the Laboratory of General Chemical Analysis, the Laboratory for Mining and Metallurgy, and the Laboratory for Industrial Chemistry. The Laboratory of Physics and Mechanics would house implements necessary for the study of physical processes. The strengths of materials; the flow of air, water, and light; the power of machinery, all required special apparatus and a separate area in which to store the instruments. Rogers envisioned rooms filled with microscopes, barometers, thermometers, hygrometers, dynamometers, burners, lamps, and even "a room fitted up for photometry" in which students could learn to measure light as produced by such materials as gases. The Laboratory of General Chemical Analysis would provide for the qualitative and quantitative analysis of organic and inorganic materials. At least two years of courses in basic, or, in his words, "general," science at the Institute would be required as preparation for work in this laboratory. The Laboratory for Mining and Metallurgy, affiliated with the General Chemical laboratory, would emphasize practical mineralogy, or "the chemical valuation of ores, and the operations of smelting and other processes for the separation and refining of materials." Students of mining and metallurgy would learn to discriminate rocks and minerals by way of "mechanical and chemical tests." Rogers coupled the laboratory exercises with instruction on models of mining and examples of equipment used in the extraction of earthen materials. In addition to the General Chemical laboratory, Rogers proposed the Laboratory for Industrial Chemistry. "The more important chemical arts and manufactures" would be the focus in this department. Students would follow similar processes illustrated in other departments but with different materials. Rather than unrefined ores or basic organic or inorganic substances, industrial chemistry called for "dyestuffs, mordants, discharges and other substances used in the operations of dyeing, color printing, and bleaching."[39]

All four of the laboratories shared both the common goal of practical instruction for regular students and a research mission. If commissioned by the MIT government or its branches represented by the Society of Arts and the Museum of Arts, the laboratories "will be used for the prosecution of experiments and investigations, . . . examination and testing of new machines and processes, and the conducting of original research." Advanced students, Rogers believed, would benefit by assisting in such research led by faculty or others.[40]

During the period in which Rogers prepared the *Scope and Plan,* he brought together, sifted, and refined all of the ideas about technical institutes he had collected over the years. Many of them came from Europe, especially France but also Germany. The homefront had less to offer. The use of the laboratory for student instruction had not been given serious attention within American higher learning. Some medical schools, colleges, and institutes allowed for limited experiences with expensive apparatus, but the costs for most institutions not founded for the purpose of science proved to be prohibitive. Rogers looked abroad once more and collected plans from programs he found compatible with the useful arts. Building on what he knew of French scientific studies, William corresponded with Henry for information on programs in Great Britain. "Can you get me any drawings and descriptions," he asked, "of the interior of the Technology department at Edinburgh, and the School of Mines, Jermyn Street? All information of a specific kind relating to the fitting up and working of practical laboratories . . . will be of great value to me." Henry responded by sending his brother plans of instruction and other materials about the Kensington Museum and the School of Mines.[41]

After completing his final survey of institutions abroad, Rogers put the finishing touches on the *Scope and Plan*. A final draft emerged in time for the annual meeting of the MIT government held on May 30, 1864. The governing body approved the document at the gathering, making it the Institute's first curriculum.[42]

In part to alleviate his failing health and in part to satisfy his continued interest in science instruction abroad, Rogers set sail for Europe a few days after MIT accepted his *Scope and Plan*. He had two educational goals while on his tour: to collect and inspect. He wanted to gather models while in Britain and the Continent for use in the Institute's varied program areas. He looked for models that he hoped would aid in the instruction of students at MIT. Students, he contended, should learn from models of "machinery, or bridges, roofs, arches and other works of civil construction and architecture." Aware of this plan, the Institute's governing board granted him one thousand dollars to purchase equipment. Rogers also left for Europe with a plan to inspect "the recent and best arrangements for working-laboratories and lecture-rooms." After visiting several places of interest, he found useful the organization of "Archer's Museum" and the Kensington Museum but left Europe uninspired by its laboratory arrangements. "As for laboratories and lecture-rooms are concerned," he remarked about his tour, "I believe we have little to learn either in England or Paris." Thus, in the fall of 1864, following his expedition, he turned his attention to assembling a faculty for the first day of classes.[43]

⌐⌐

Rogers set February 1, 1865, as the date MIT would open its doors to students and launch the plan of instruction. While the war still worked against the prospects of opening the Institute, Union victories had turned the tide clearly against the rebellion. The idea of MIT could claim more attention in this new context than was previously possible. To meet the goal he had established, Rogers began to recruit faculty. For the chair of mathematics he selected John D. Runkle. A graduate of Harvard's Lawrence Scientific School, former pupil of Benjamin Peirce, and affiliate of the *Mathematical Monthly,* Runkle was suited to fill the theoretical needs of the useful arts mission at MIT. As an original member and early promoter of Rogers's first educational proposals before the legislature, Runkle knew well the mission of the Institute. For the professorship of mechanical engineering William Watson had a stronger applied background than that of Runkle. After graduating from Harvard College, Watson stayed with the institution as an instructor of mathematics. He left, after a time, for further study at the Ecole des Ponts et Chaussees in Paris, where he became immersed in civil engineering. After his return to Harvard as a lecturer, Rogers lured the engineer to teach at MIT. Francis H. Storer was recruited for the professorship of general chemistry, Ferdinand Bocher for French, and W. T. Carlton for drawing. Rogers, meanwhile, assumed the professorship of geology and physics, in addition to his work as president of the Institute. The humanities, it should be noted, were not overlooked. William P. Atkinson filled the professorship of English Language and Literature, and in later years George Howison taught Philosophy of Science and Logic. With this collection of faculty, Rogers could argue that the Institute had addressed calls for modernizing American higher learning.[44]

Although he missed his original goal by nineteen days, Rogers wrote in his diary for February 20, 1865: "Organized the School! Fifteen students entered. May not this prove a memorable day." The oft-cited entry refers to the first classes held at a rented space in the Mercantile Building in downtown Boston. Construction on the Back Bay lands would take another year before Rogers could hold his lectures and laboratory instruction on the lots. In the meantime the Institute held a preliminary session for students desiring preparation for its official opening expected in the fall. When the fall session arrived, Rogers greeted seventy students and five additional faculty members.[45]

Of the five new professors, Rogers had taken a special interest in recruiting Charles W. Eliot. After having studied and held an instructorship at Harvard, Eliot went to Europe during the Civil War to further his training in chemistry. While abroad, he visited all the "great and well-organized Polytechnic Schools," including those of "Paris, Karlsruhe, Stuttgart, Zurich, [and] Vienna." He worked in the famous Kolbe laboratory toward the end of his tour and acquired an interest in the useful arts. Eliot learned of the great advantages of European manufacturers in having the support of such institutes and the "difficulty for American manufactures" who lacked similar support. "Science, whether pure or applied," he remarked, "is not yet naturalized in the United States. . . . when the American people are convinced that they require more competent chemists, engineers, artists, architects than they have now, they will somehow establish the institutions to train them." Rogers and the Institute's faculty were instructing their preliminary students for precisely these ends when Eliot made his observations. Eliot's years in Europe had left him largely unfamiliar with the emergence of MIT. Although he had maintained correspondence during his tour abroad with his old friend Francis H. Storer, who had become a professor there, Eliot's attention was directed elsewhere. One of the leading textile companies in New England had sent him an offer that most believed he could not refuse: the superintendency of a mill factory in Lowell, Massachusetts, a five thousand—dollar annual salary, and rent-free housing. On the surface the offer agreed with Eliot's taste for administration. If he chose to make a start in the manufacturing world, few better offers would come his way. On the other hand, accepting the invitation would likely mark the end of his academic career, in which he had invested the previous decade and his time in Europe.[46]

To Rogers's surprise and relief, Eliot turned down the superintendency. Upon hearing of Eliot's decision, Rogers immediately dispatched a letter to Europe, offering him a chemistry professorship at the Institute. With the fall session, beginning October 1865, a mere four months away, it might have seemed like a long shot. From Rogers's point of view, however, Eliot matched the Institute's useful arts mission in ways that many others could not. "My great anxiety now," he wrote to Eliot in June 1865, "is to make a good faculty of instruction, and I want you to be one of the number." Aware of the friendship between Storer and Eliot, he suggested that the two of them could share the duties of teaching in the chemistry department and could work out between them what branches, general or industrial, each would direct. Eliot found the offer intriguing, but, because of his lack of familiarity with MIT, he requested more information about the Institute's founding principles. Rogers and Storer responded by showering Eliot with information and "pamphlets," which likely included the *Objects and Plan* and the *Scope and Plan*. Rogers emphasized the secured endowment, progress on the building located on the Back Bay, and plans under way for a comprehensive "working laboratory." In addition, Rogers made clear the kind of administrative approach adopted by the Institute. "Long experience has taught me," he informed Eliot, "the importance of giving to each professor a wide latitude in the choice and use of his plans and means of instruction, making him, in fact, within reasonable limits, the sovereign in his department." Rogers's letter persuaded Eliot that they shared mutual scientific and educational interests, for in late July 1865 Eliot sent an acceptance letter from Paris. Eliot needed no further coaxing; he called Rogers's points "very satisfactory" and described himself as contented and gladdened by the opportunity.[47]

With Eliot's acceptance the chemistry department felt complete to Rogers. Following his efforts in recruiting faculty, however, came concerns over construction on the Back Bay. Delays caused Institute faculty and students to return in the fall to rented spaces in the Mercantile Building area. The spring term of 1866 fared slightly better. The main MIT building on Boylston Street by then had walls, a roof, and two finished rooms. One lecture hall and a laboratory could be put to use, but the rest needed finishing. After a year of delays, Rogers finally moved all functions of the Institute to the Back Bay in time for the fall term of 1866. In the basement of what came to be known later as the Rogers Building, students practiced exercises in the laboratories for general, geological, and industrial chemistry. The first floor housed the president's office, lecture rooms, and the physics laboratory, in addition to a meeting room and office space. Lectures in mathematics, civil engineering, modern languages, and astronomy, were delivered on the second floor. A "half story floor," between the second and third, provided for museums and library space, lecture and modeling rooms, and two faculty studies. The third floor was dedicated to drawing, modeling and lectures rooms for architecture, mechanical engineering, and mathematics as well as additional office spaces. The fourth and final floor held faculty offices, a photographic laboratory, and a freehand drawing room.[48]

Characteristic of Rogers's useful arts ideal, the kind of spaces created for MIT reflected a dualism between theory and practice. The exterior of the building, elegant and classical, alluded to the theoretical aims of the Institute. Its facade, with four classical columns over the main entrance, drew on the imagery of antiquity and would have called to mind knowledge and scholarship of transcendent value. The interior, particularly the metallurgical laboratory, was radically stark, by contrast. On plain brick surfaces stood wooden boxes filled with supplies, next to an array of pipes, tools, vents, and furnaces that resembled the floor of an industrial factory. To Rogers's mind, however, the difference between the Institute and a common factory was that factories produced commodities, whereas the Institute sought the production and advancement of knowledge.[49]

For two years Rogers took pleasure in seeing the useful arts principles in the *Objects and Plan* and *Scope and Plan* come to fruition. But the challenges he had experienced in reaching that point were not without costs. At a faculty meeting at the start of the fall term of 1868, Rogers began to feel uncomfortably hot. Dismissing the hot flash at first, he soon felt giddy and faint. Nevertheless, he continued at the meeting until paralysis struck the left side of his face and, in midsentence, found his "articulation . . . oddly obstructed." No one expected what was to follow.[50]

CHAPTER EIGHT

Reception of the Idea

EMMA CALLED IT "nervous exhaustion." MIT's mathematics professor John D. Runkle dubbed Rogers's illness "Institute on the brain." Rogers himself said he was "liable to much nervous perturbation." When the stroke hit, leaving him with partial but temporary paralysis, he broke off all engagements in Boston and decided to stay with his brother Robert in Philadelphia. From fall 1868 to summer 1871 William spent his winters with his brother and his summers on the coast in Newport, Rhode Island. The time away must have affected him for the better, for by 1871 he began spending his winters in Boston, while continuing to rest in Newport during the summer. He and Emma enjoyed Newport enough to build a summer home there, which they called "Morningside." His visits to the coast offered some relief from demands in Boston where obvious loads on his health had included the war, scientific and professional appointments, and the founding of MIT.[1]

Although Rogers's sudden decline forced him to leave behind his official commitments at MIT, he followed and kept a hand in significant decisions made in his absence. Since its inception, he'd held the Institute's first presidency as well as the professorships of geology and physics. In the summer before the onset of his illness he hired an assistant professor to take over the physics duties. When the stroke left him, as Emma put it, "unable to walk more than a few steps in his room, to read or to listen to reading, or to do any mental work," the Institute responded by granting Rogers a leave of absence. His friend Runkle served as MIT's interim president for the first eighteen months. When Rogers officially resigned, Runkle continued as MIT's second president until 1878. All the while, during this period of illness and transition, the Institute faced fundamental questions about its future, which made their way back to Rogers's bedside.[2]

The dilemmas raised during MIT's formative period cannot be divorced from the general reception of technological institutes in the 1860s and 1870s. Public discourse over the idea of MIT and like-minded establishments germinated from two long-standing academic feuds. Within one forum of debate, advocates of culture continued to spar with scientists over the curriculum. In the other, heated exchanges intensified between members of the science community over defining the proper approach to scientific studies. The pages of popular and scientific periodicals were increasingly consumed by both feuds. "In the matter of 'technical education,' which now forms a prominent topic of discussion on both sides of the Atlantic," protested the *Scientific American,* "there has been, hitherto, altogether too much talking." Despite such protestations, how and where college students studied science and technology commanded a growing share of the public interest.[3]

The founding of institutes such as MIT rekindled hot partisan embers that had glowed since the appearance of the Yale Report in 1828. For many culture advocates of the 1860s and 1870s the report outlined essential arguments for maintaining a curriculum based on the ancient languages that applied in part or in whole to the developments of the postbellum era. Yale president Noah Porter led among those who reintroduced the principal ideas of the report. "We assert that the study of the classical languages," he reminded colleagues, "should be universally preferred to any other as a means of discipline in every course of liberal education." He agreed with "the theory of education . . . that certain studies (among which the classics and mathematics are prominent) are best fitted to prepare a man for the most efficient and successful discharge of public duty." In his view the classics disciplined (that is, exercised the mind) and offered the best of human culture (that is, literature and "knowledge of man") more effectively than any other form of liberal studies. As such, the traditional curriculum prepared students for the "commanding position" deserved by the "thoroughly cultured man." The increasing attention to scientific and practical studies as well as to research at the expense of teaching had all weakened the college's ability to serve its traditional cultural mission. Faculty, according to Porter, should pay nearly exclusive attention to instruction over original investigations. "No mistake can be greater," he emphasized, "than to suppose that a college gains very largely by adding to its corps of professors eminent personages who have little or no concern with the business of instruction, or who come rarely in contact with the students." Porter hoped that by restating the Yale Report's ideals he could slow the changes then occurring in mid-nineteenth-century American higher education.[4]

The rise of schools of science, institutes of technology, and the emergence of the modern university certainly figured into Porter's concerns. Sometime in the distant future, he speculated, universities may indeed be "desirable and possible with enterprise, patience, time, and energy." But, as he saw it, the 1860s and 1870s were not an opportune time for such an undertaking in America.[5]

Porter hardly stood alone in expressing these convictions. Other culture advocates joined in defending the classics as the bedrock of college-level education. George Park Fisher, a professor at Yale's Divinity School and editor of the *New Englander,* considered Latin and Greek the basis of a true undergraduate course of studies. "The study of the classical languages and literature," he insisted, "is a leading, essential, indispensable part of such a scheme of education." T. W. Higginson, a Massachusetts Unitarian minister and contributor to the *Atlantic Monthly,* repeated the same in a "plea for culture." What these languages offered was a buffer against "the strong tendency to make all American education hasty and superficial." His idea of culture meant, at bottom, "the training and finishing of the whole man, until he sees physical demands to be secondary, and pursues science and art as objects of intrinsic worth." He made clear that practical or useful studies obfuscated higher education's cultural mission.[6]

Some culture advocates, such as James Jackson Jarves and Paul Ansel Chadbourne, faulted science for leading American higher education astray. Jarves, a well-known art collector and author, questioned the impact of philanthropy that aided the new education at the expense of the old. "Rich men contribute liberally to support a college, institute of technology, or museum of natural history," he ruminated, "on the general principle of their usefulness . . . without comprehending specifically anything of their studies or doctrines." He called on philanthropists to redirect their resources toward museums of art rather than museums of natural history. Like Jarves, University of Wisconsin president Paul Ansel Chadbourne raised concerns about financing scientific and technological education. "We are losing vastly, absolutely wasting our means," decried Chadbourne, "especially in our attempts at industrial education, while so many colleges are attempting to teach everything without having the means of thoroughly teaching anything." His solution to the problem was division of labor. Institutes of technology should meet the demand for practical education without concerning themselves with "liberal provisions for college or scholastic studies." Colleges, on the other hand, should restrict their offerings to the traditional curriculum for those "who are seeking knowledge for some other purpose than as mere multipliers of dollars and cents." Chadbourne urged that the division of labor would keep the traditional course from becoming distracted by practical studies.[7]

In large measure Porter and the circle of culture advocates directed their statements, all of which appeared in the late 1860s and early 1870s, to general movements in higher education. While they may have cited specific examples, their main point was to defend the merits of the classical curriculum and call attention to the increasing emphasis on science, or certain forms of science, over the study of culture.

One of the most lively points of contention between culture advocates and scientists, however, originated with the founding of MIT itself.

MIT's founding and establishment in the 1860s gave Jacob Bigelow and William P. Atkinson an occasion for a series of addresses related to the idea of institutes of technology. The addresses, whatever their accuracy in representing the values that had led to the creation of MIT, played a central role in the Institute's initial reception. Their ideals and aspirations as well as their concerns and provocations became fused with the institution, at least in public discourse, and influenced public perceptions of MIT. When Bigelow spoke at the Institute, he was hardly an unknown figure to Boston-Cambridge intellectuals. He had led a successful private practice as a physician, authored several studies in botany, worked as a part-time chemist, and held the professorship of *materia medica* at Harvard's Medical School. When, in 1816, Harvard established the Rumford Professorship on the Application of Science to Useful Arts, they turned to Bigelow to fill the position. His interests in both practical and theoretical sciences matched the requirements of the professorship. By the time he turned his attention to MIT, he had served for over fifteen years as president of the American Academy of Arts and Sciences. The younger, lesser-known Atkinson organized his thoughts about the Institute from the perspective of a humanist. After receiving an A.M. degree from Harvard, Atkinson was drawn to the idea of MIT because of its attention to modern, as opposed to classical, humanistic studies. The institution's interest in experimenting with instructional methods also attracted him. "I have had much to say," he remarked about the recitation method, "against the absurdity of learning by rote that which should be reserved till it can be intelligently grasped and comprehended." For Bigelow and Atkinson the Institute offered a corrective to the traditional collegiate model. Yet their approach to promoting the idea differed as greatly as their backgrounds.[8]

Bigelow had a polarizing effect on opinion makers when he gave a pair of addresses on MIT's founding in 1865. In the first address, *On the Limits of Education. Read before the Massachusetts Institute of Technology, November 16, 1865,* he emphasized that a knowledge explosion had occurred in the first half of the nineteenth century. As a result of this expansion in knowledge, particularly in areas of science and technology, Bigelow declared the era of the "general scholar" as finished. "No individual," he stated, "can expect to grasp in the limits of a lifetime even an elementary knowledge of the many provinces of old learning, augmented as they now are by the vast annexation of modern discovery." Given the revolution in intellectual life, he considered it unreasonable to devote four to five years of collegiate study to the classical curriculum. The old college brought students to "the many doors of the temple of knowledge, without effecting an entrance to any of them." The purpose of MIT, as he understood it, was to address this explosion of knowledge with a course of study that allowed for specialization and the exploration of useful knowledge.[9]

Bigelow's second address, *On Classical and Utilitarian Studies,* spoke directly to the merits (or demerits) of the classical curriculum. Defining education as a combination of mental discipline and useful knowledge, he declared ancient studies as inadequate on both counts. Antiquity had failed to produce "any solid and lasting good" because the culture had been "extensively given to fictions, words, and profitless abstractions, rather than to the augmentation of permanent knowledge." The modern era, by contrast, had expanded knowledge for social and material utility, for improving the human condition. When it came to mental discipline and useful knowledge, he continued, any subjects of the modern age compared favorably or better to those of the ancient. While conceding that "classical literature . . . may well enter into the foundation of the most liberal forms of education," he reaffirmed his belief that the languages alone failed to address "the intellectual demands of the present."[10]

Despite his bold claims, it'd be reductive to call Bigelow's views mere anticlassicism. It's true that he used inflammatory language when describing the classics as "heathen mythology," rife with "savage attributes, brute instincts, and exceptional morality, [that] override the more modern sentiments of humanity, honor, and Christian charity." But he also acknowledged the "copious, majestic, expressive, and musical" qualities of classical literature. Holding both views conformed with his belief that the curricular landscape could support more than one model. His addresses looked to situate the MIT ideal within this landscape set in the modern world.[11]

With equal conviction, although less provocatively, Atkinson pursued the same ends. Having witnessed rancorous debates between scientists and classicists over such institutions as MIT, he sought "to place the merits of both sides of the question in dispute in a juster light than is done by the extreme partisans of either." With this goal in mind he addressed members of the Institute with a lecture on *Classical and Scientific Studies, and the Great Schools of England* (1865). His focus on English systems of education offered an indirect appraisal of midcentury American higher education. By dealing comparatively with educational problems across the Atlantic, he could raise prickly issues at home without implicating specific institutions. Drawing upon a study commissioned by the English Parliament in the early 1860s, he discussed examples of antiscience and antimodern sentiment among England's educational leadership. Administrators surveyed revealed signs of antiscience sentiment when they stated that "physical science is not taught" at their institutions or that they "hardly know what their [scientific studies] value is." Antimodern expressions revealed a similar resistance to modern languages and the fine arts. Without science or modern humanistic studies, he stated, English students received mostly Greek, Latin, and some mathematics. According to the report, students, when evaluated, showed embarrassingly little command of even these classical studies.[12]

Having described the unbalanced character of English education, Atkinson followed with what he believed were the root causes. On the one hand, advocates of science misrepresented the nature of their work to their critics. Scientists "have been too prone to confound education with information," mistaking the development of the mind with practical knowledge. "They have lost sight of that liberalizing development of all mental powers which constitute a true education, and which no mere pouring in of any amount of useful information can ever accomplish." Similarly, classicists unnecessarily restricted the range of acceptable collegiate studies. They contradicted their own belief in the broad and "symmetrical" purposes of education. Classicists themselves should concede that the means to a true liberal education "must be something of far wider application and greater power than the grammar of two dead languages." He recommended that partisans of each side should avoid mutually exclusive models and consider a balance between extremes.[13]

The Bigelow-Atkinson addresses caused a stir, and, not surprisingly, it was Bigelow rather than Atkinson who drew the sharpest responses. Some critics leveled general charges against them, while others produced detailed counterpoints to the published lectures. Culture advocates at the *New Englander*, for example, made general remarks about Bigelow's partisan spirit. They took issue with his "extreme ground in favor of giving very great prominence to the study of the sciences and the modern languages, at the expense of the classics." Critics differed with the author "upon almost every position which he takes and seeks to defend." By association MIT came under fire. "The advocates of Dr. Bigelow's system," they concluded, "are . . . one-sided men." In this sense perceptions about Bigelow and the Institute became linked.[14]

The *North American Review* ran a series of articles in the two years immediately following Bigelow's address that continued along the same lines, offering specific rebuttals to the idea of the Institute. The *Review*'s commentators contended that "what Dr. Bigelow means by Education is not education at all but elementary instruction." To them he had misconstrued higher learning to mean "the imparting of information useful for some purpose other than education." Information transfer is not the point of a true education, they responded. Colleges should aim toward a higher goal that would encourage students to ask why something is done rather than how. That Bigelow held such misguided ideas, they maintained, revealed the need for scientists and technologists to receive a liberal, classical education. The idea of MIT thus fell under similar scrutiny for having a potentially ill-conceived curriculum: "If it is men and not machines, that are to be turned out by the new Institute the methods and the test of success will have to be something beyond a rigid adaptation to the performances of particular tasks." The MIT addresses, as such, raised concerns over what kind of graduates the program would produce.[15]

In contrast to the attention given to Bigelow, Atkinson received fewer and less heated replies. Atkinson's address nevertheless also raised eyebrows and thus figured into MIT's reception. Critical commentary from the *North American Review* challenged his claims that England's overemphasis on the classics could inform America's educational dilemmas. The failure of the English to educate their students fully in the classics, argued the *Review,* spoke volumes about English instruction and little about classical education. While conceding that science might teach certain powers of observation, the *Review* maintained that no empirical evidence existed to confirm that this indeed was so. Using the principles of science against Atkinson, the commentary remarked that "this ought to be shown in detail and by experiment. . . . What we want to know is the comparative effect of different studies upon the same faculties." The tenor of the remarks matched the tone of Atkinson's address. Whereas Bigelow's abrasive rhetoric generated equally abrasive replies, Atkinson had attempted a balanced review of classical and scientific education, drawing a like response from critics. In the end, however, the *Review* disagreed with Atkinson's conclusion that science may have an equal educational value to that of the classics. Because no "experiment" had studied the effects of classical or scientific education on students, its authors left open the possibility that both may offer "highly important" modes of mental activity. Until such experiments conclude otherwise, the argument continued, the classical curriculum should remain undisturbed as the basis for American higher education.[16]

From William Barton Rogers's perspective the idea of MIT had taken on a life and meaning of its own independent of the original intent. Rogers regretted, in fact, the impact and distortions created by the inaugural lectures at the Institute. "The recent discussion here and elsewhere," he reflected, "on the relative value of scientific and classical culture in our schools and universities seem to threaten an antagonism which has no proper foundation in experience or philosophy." When E. L. Youmans, popularizer of science and an itinerant science lecturer, offered to deliver a talk at MIT on "the superiority of the sciences over the classics," Rogers had had enough. Youmans, who was then editing a volume of essays by leading British scientists on the merits of scientific education, wanted to talk about how the classics offered the "lowest kind" of mental discipline, that "advocates of the dead languages have failed to prove" their worth. His point was that the languages promoted passive learning, as in the uncritical acceptance of ancient knowledge, while science stimulated the faculties of observation and independent judgment.[17]

Rogers turned down the offer and described to Youmans the spirit with which the Institute had been founded. "Some advocates of the Old System," explained Rogers,

> are trying to make the impression that friends of progress in education are as a matter of course the *enemies* of classical studies—while as you know we would have such studies not *excluded* but only subordinated in a complete curriculum of training and instruction. The intellectual and aesthetic discipline obtained in the *Study of Languages, Modern* as well as ancient, is of undoubted value and might be provided for in every comprehensive course of education. But this training can in no degree replace the invigorating exercise of the observing and logical faculties so peculiarly the function of scientific studies. Let the classics have their place among the instruments of intellectual culture, but in general education let them be kept within the modest limits appropriate to them, in which they shall not as they now so often do stand in the way of the broader, higher, and more practical instruction and discipline of the natural, mental, and social sciences.[18]

Youmans's offer to lecture at the Institute represented a current of public perceptions about MIT. The Institute, some assumed, was a bastion of anticlassicism. In such instances Rogers attempted to correct misperceptions by reminding observers of MIT's original intent. According to him, who devoted the institution's founding documents—such as *Objects and Plan* (1860), *Scope and Plan* (1864), and the inaugural *Catalog* (1865)—to the useful arts ideal, the Institute aimed toward both breadth and depth in science instruction. Most institutions of the era addressed either one or the other but rarely both in the same institution. Technical and engineering training institutes (such as West Point or Rensselaer) served largely practical interests, while the Scientific Schools (such as Harvard's Lawrence or Yale's Sheffield) emphasized nonpractical areas. The attempt to bring breadth and depth under one roof reflected Rogers's own style of research, which contributed to and made sharp distinctions between practical and theoretical knowledge. His early work as director of the Virginia Geological Survey in the 1830s and 1840s (largely practical) and later debates over evolution with Harvard zoologist and geologist Agassiz in 1860s Boston (largely theoretical) grounded his ideas about the nature of the Institute and its purpose in American higher education.[19]

Rogers's intentions, therefore, were not to antagonize classicists. Far from it, for he had received training in the ancient languages as a student at William and Mary and lauded the benefits gained from humanistic studies broadly conceived. At the Institute the humanities (modern languages, history, philosophy), while "subordinated," nevertheless accounted for a substantive part of the original curriculum as well. Rogers believed a broad, deep, and humanities-infused program was vital to the advancement of the scientific community in the United States and for preparing of the next generation of scientists. "The most truly practical education even in an industrial point of view," he stated, "is one founded on a thorough knowledge of scientific laws and principles, and which unites with habits of close observation and exact reasoning . . . the highest grade of scientific culture would not be too high as a preparation for the labors of the mechanic and manufacturer." While Rogers might have attempted to correct misconceptions, the addresses and the responses they elicited attracted far more attention and shaped public opinion about the new institution.[20]

The crux of the classicists' position centered on the belief that the ancient languages should continue as the dominant force in undergraduate studies. They believed that the time wasn't yet ripe for financing the expansion of scientific instruction. If anything, science education posed a serious distraction to the true cultural purposes of American higher education. Vocal critics of institutes of technology often interpreted such institutions as a threat to the classical college ideal, rather than as a complementary form of higher learning. Bigelow's outspoken denunciation of Greek and Latin studies likely encouraged the perception that MIT's founding represented the fruits of an anticlassicist movement. Science supporters of the 1860s and 1870s spent a great deal of energy responding to the varied criticisms from culture advocates; just as important, however, they also engaged in squabbles among themselves over proper modes of science instruction. Differing views about separate schools of science and technological institutes, in particular, divided the scientific community. Their debates centered on the degree to which higher education should emphasize abstract versus practical scientific studies.

As for scientists who responded to classicists, the knowledge explosion argument provided reformers with a broad-based rationale for modifying the undergraduate course of study. With a tone of impatience the editor of the *Scientific American,* for example, responded directly to those classicists who argued that the time was not ripe for change. The impatience, in this case, stemmed from the assumption that the entire discourse against scientific and technological education delayed an inevitable march toward progress. "If we are willing to look and wait," argued the editor, "till the philosophers have ceased to wrangle on this subject and have come to an agreement among themselves, the day of judgment will certainly dawn on an earth unprovided with technical institutions." With the founding of technological institutes construed as inevitable, the bickering over whether or not to have them "is of minor importance." The editor urged immediate action for higher educational reform.[21]

While calling attention to the expanding base of knowledge, the *Scientific American,* when presenting the topic of "Scientific versus Classical Education," conceded that some study in the classical languages could help prepare those interested in science. A few months of Latin and Greek, at best, would serve the purpose. Four years of ancient languages, on the other hand, would make sense only "if our lives lasted a thousand years. . . . But at the present time, it can only be acquired at the expense of other information." No longer could the classical curriculum suit the needs of scientists, for "the accumulation of knowledge" demanded that "a choice must be made between different kinds of learning." Students faced a choice between ancient languages, with their myths and "delusions," and the "positive and accurate knowledge" assembled by scientists. In chemistry alone, argued the *Scientific American,* the field is "enormous" and "constantly spreading." For these reasons the editor advocated a college-level reform with two specific points: the classics could still play a role in the college admissions process, but colleges themselves should abandon the teaching of Latin and Greek entirely. Reformers with such beliefs insisted that only change of this kind would overcome the "trammels of this prejudice," which called the classics "the most valuable knowledge of all." The "rational" and "proper" course for higher education would be to teach science through the use of instruments and experiments.[22]

Much like the *Scientific American,* the *Manufacturer and Builder* also invoked the knowledge explosion argument when calling into question the study of Latin and Greek. Although the ancient languages were once considered "all-sufficient," the age of industry required a change in such thinking. Reminiscent of Bigelow's challenge to classicists, the *Manufacturer and Builder* cited that "the ancients were by no means the best physicians, having generally the most absurd notions about anatomy, physiology, pathology, materia medica, etc." For physicians to spend time on the classics would violate "common sense." The preparation of scientists, and physicians in particular, should, according to the periodical, avoid "the absurdity of the ordinary course of study" and focus on "the superior result of such training in which observation and experience are the basis." Rather than depicting the dilemma as a choice between scientific and classical education, culture critics of this view cast the conflict as one between "common sense versus classical education."[23]

In addition to claims about an expanding knowledge base, the *Manufacturer and Builder* used nationalism to defend the appropriateness of scientific studies with regard to its timing and financing. The argument invoked nationalistic arguments that used nativism to fuel interest in the promotion of science in colleges and universities. "If we ever intend to become independent of foreign-skilled labor" remarked the publication, "some more strenuous means must be adopted to afford a proper mechanical and technical education to the young men." Science education, in this light, served the politics of independence and national competition in an age of industrialization. "There is no reason why educated foreigners should be induced to emigrate by the offer of enormous salaries to superintend our manufacturers," continued the argument, "when these positions might and should be held by our own country." Citing the state-supported polytechnic universities of Germany, scientists could claim that in America "we want, and must have, such institutions . . . schools free if possible, and which will open their doors to the youth who, though poor, thirsts to that knowledge." Only then, the *Manufacturer and Builder* suggested, could the nation compete with the influx of highly trained scientists, technologists, and artisans from Europe.[24]

When not engaging in debates with classicists, some scientists began to defend the need for applied studies to their colleagues who favored abstract science. Prominent scientists of the period, men such as Harvard's Louis Agassiz and University of California geologist Joseph LeConte, made this a difficult task. Agassiz rejected the "practical" bent of the MIT ideal and used his prestige to oppose efforts to introduce similar tendencies at Harvard's Lawrence School. LeConte also vigorously challenged those who supported applied studies as a means to enhance "material comfort and happiness." Rather, "the highest end of science is not to lead us downward to art, but upward to the fountain of all wisdom." This tension within the scientific community led to a rift between science school advocates and technological institute supporters, although in each kind of institution both applied and abstract branches of science were represented. Practical science advocates employed the same arguments used against classicists—the inadequacy of existing systems, nationalism, and the knowledge explosion—to sway their colleagues.[25]

On the inadequacies of existing American science instruction, the leading statement of this debate during the 1860s and 1870s came from Charles W. Eliot, then a professor at MIT. Eliot's well-known "New Education" article in the *Atlantic Monthly* of 1869 gave several reasons why schools of science and parallel courses would never satisfy the needs of the scientific community. Most important to him, science schools and parallel courses could not escape their "ugly duckling" role on American campuses. The tried experiments suffered from an inferior position with regard to "property, numbers, and confidence of the community." Enrollments remained low, observed Eliot, at the most prestigious science programs as a result. He referred to low entrance requirements, low graduation requirements, and an uncoordinated curriculum as additional reasons for the absence of progress in the separate schools as well as in parallel courses. The solution for science in higher learning, as far as Eliot was concerned, rested in the idea of MIT and similar technological institutes. In such independently established organizations, he believed, the freedom to pursue the pure and applied sciences would be left unhindered by traditions or inferior status.[26]

Other advocates of the Institute ideal focused on nationalism and knowledge expansion to further their claims. Charles G. Leland, founder, editor, and contributor to the *Continental Monthly,* supported the idea of MIT, as presented in Rogers's *Objects and Plans,* because it offered an alternative to science instruction of "the old literary regime." Leland had developed an understanding of educational reforms occurring abroad during his early years, having studied in Heidelberg and Munich and having lived in Paris before graduating from Princeton. Believing that "the time has come for the establishment of such Institutes," Leland argued for the need to compete with European science and industry that benefited from such programs. While the whole of American higher education languished under a "feudal" order, the idea of MIT offered "the application of generous, intelligent culture to practical pursuits—the whole to be based on exact science." Leland also looked to institutes to deal with the knowledge explosion in science. "The growth of science," he remarked, "has . . . so vastly increased, that the proposition to reform the old system of study is really not to tear it down, but to build it up, to extend it and develop it on a grand scale." Disillusioned with separate schools such as Harvard's Lawrence School, critics looked to places like MIT to undertake the study of science and "technological information of every kind" with equal vigor. Leland urged readers to consider MIT as the first "thoroughly *scientific* university" in America.[27]

Returning to the inadequacies of existing science instruction, Horace E. Scudder, a science enthusiast and manuscript editor for several presses, wrote on the merits of "Education by Hand" for *Harper's Monthly* readers. He was concerned about the dislocation of workers in an age transformed by industrialization. Scudder surveyed the decline of the apprenticeship system and the potential for technological institutes to fill the void. The process of industrialization and "other changes in our more complex society," he explained, have rendered the apprenticeship system obsolete. MIT and similar institutes offered an "education of the hand . . . with the education of the mind" for those interested in the industrial and scientific branches of knowledge. For would-be apprentices the Institute provided experience with tools and instruments that could be found both in the machine shop and the scientist's laboratory. For aspiring scientists, he continued, the program gave a theoretical foundation for the pure and applied sciences.[28]

Not all reformers believed that MIT could address the practical and theoretical needs of the scientific community. Some, like Edward Atkinson, found MIT's course of study suited for advanced students of science but not as appropriate for the would-be laborer or mechanic. Atkinson was an industrialist and economist who had started his career as a textile factory floor-sweep at the age of fifteen. To him, a fervent MIT supporter, the instruction offered by the Institute for those interested in the mechanic arts was "suitable for a graduate of a [college preparatory] grammar school." For others who lacked such preparation, another system would be needed to advance "primary instruction in the use of the hand." Many agreed with the kind of observations made by Atkinson, observations that led to the founding of applied studies programs. New Jersey's Stevens Institute, established in 1867, directed its focus to "the boy who has a positive talent" in the mechanic arts. According to those who visited the institution, few could match "the same facilities as the Stevens Institute" for practical studies. The Worcester Free Institute, which appeared in the 1870s, was designed for the same purpose: to address the demand in applied courses of study. "The distinguishing feature of the school," applauded practical studies advocates, "is the practical part of the course of mechanical instruction." The machine shop used by students followed a structured, formalized apprenticeship. With its practical emphasis the Worcester Free Institute was "working out one of the most important questions of technical education . . . namely, whether theoretical and practical instruction can, to any great extent, be successfully combined in a technical school." Cornell University joined the applied science fervor when it created a program of civil engineering designed to meet the practical needs of industry and science. Together with the "numerous special schools of technology" founded in the 1860s and 1870s, Cornell's program, its supporters claimed, helped staunch the tide of "students seeking instruction in the higher branches of applied science" who would have left for "the polytechnic schools of Germany or France."[29]

Pressed between abstract and applied factions within the scientific community, MIT faced a dilemma. Science schools claimed priority to abstract science, and emerging programs of applied science offered exclusive attention to practical studies. The idea of MIT, which aimed to offer both branches of science, faced criticism from the most abstract- as well as practice-oriented scientists. Observers waited to see whether MIT could fulfill the needs of both theory and practice within one institution.

❧❧

While the feuds between classicists and scientists as well as those within the scientific community provided a general indication of how technological institutes were received, developments at MIT offered a case study. The status of MIT's stu-

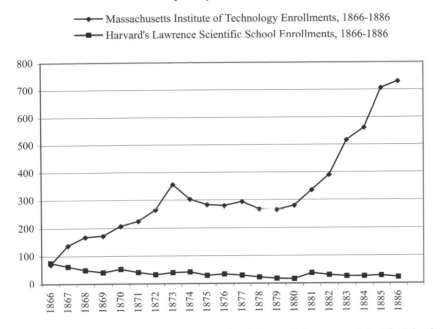

—◆— Massachusetts Institute of Technology Enrollments, 1866-1886
—■— Harvard's Lawrence Scientific School Enrollments, 1866-1886

Fig. 8.1. Comparison of enrollments at MIT and Harvard's Lawrence Scientific School, 1866–1886. These figures are based on a review of the enrollment figures listed in MIT's *Course Catalog* and Harvard's *Presidential Reports* for the dates surveyed. MIT enrollments were: 1866, 69; 1867, 137; 1868, 167; 1869, 172; 1870, 206; 1871, 224; 1872, 264; 1873, 356; 1874, 303; 1875, 283; 1876, 280; 1877, 293; 1878, 267; 1879, 264; 1880, 279; 1881, 335; 1882, 390; 1883, 516; 1884, 561; 1885, 706; 1886, 730. Harvard enrollments at the Lawrence School of Science were: 1866, 75; 1867, 61; 1868, 49; 1869, 41; 1870, 52; 1871, 41; 1872, 32; 1873, 40; 1874, 42; 1875, 29; 1876, 34; 1877, 29; 1878, 22; 1879, 17; 1880, 16; 1881, 37; 1882, 30; 1883, 25; 1884, 26; 1885, 28; 1886, 22.

dent body, curriculum, and relationship to Harvard generated basic measures of the Institute's reception.

During most of William Barton Rogers's absence in the late 1860s and 1870s, the Institute drew large numbers of students who would have sought scientific studies in other programs in the United States or Europe. A mere three years after its founding, MIT had three times the number of students enrolled in Harvard's Lawrence Scientific School. The Institute's enrollment numbers steadily increased, widening the gap between the two schools, until the economic panic of 1873. The panic and ensuing depression slowed the gains they had made and for a time preoccupied the schools' officials. When the economic downturn passed, however, so, too, did the cause for their concern.[30] Initially, students came for the most part from within

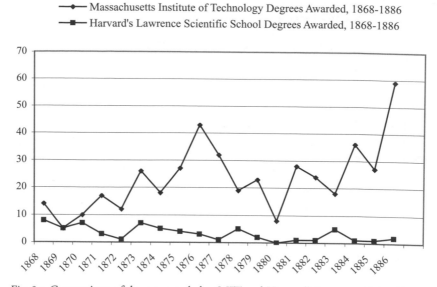

Fig. 8.2. Comparison of degree awarded at MIT and Harvard's Lawrence Scientific School, 1868–1886. Records for degrees awarded can be found in MIT's *Course Catalog* and Harvard's *Presidential Reports* for the dates surveyed. By year MIT degrees awarded were: 1868, 14; 1869, 5 1870, 10; 1871, 17; 1872, 12; 1873, 26; 1874, 18; 1875, 27; 1876, 43; 1877, 32; 1878, 19; 1879, 23; 1880, 8; 1881, 28; 1882, 24; 1883, 18; 1884, 36; 1885, 27; 1886, 59. By year Harvard Lawrence School of Science degrees awarded were: 1868, 8; 1869, 5; 1870, 7; 1871, 3; 1872, 1; 1873, 7; 1874, 5; 1875, 4; 1876, 3; 1877, 1; 1878, 5; 1879, 2; 1880, 0; 1881, 1; 1882, 1; 1883, 5; 1884, 1; 1885, 1; 1886, 2.

the state. As the Institute became more widely known, the population broadened and attracted those from regional, national, and international locales. During the first five years the number of students from outside of Massachusetts rose from five to forty-three. The total enrollment figures included regular and special students. Students who had an interest in science but perhaps did not want to complete a full course of study were drawn to MIT.[31]

The actual number of those who enrolled and those who finished a program of study varied from year to year, and substantive differences existed between the Institute and Harvard's Lawrence School in this regard as well. From the start Rogers had an idea of what kind of student would be successful at the Institute. His experiences with students at William and Mary and the University of Virginia would have reminded him that student behavior had a central role to play in the operations of any institution. Naturally, this meant that he wanted MIT students who would be "amiable and correct" in behavior. The Institute could not afford any reckless behavior given the expense of laboratory equipment and other apparatus used by students at MIT. Good behavior, however, was hardly enough, for students also needed to have a "decided aptitude and taste for scientific studies" in addition to a "sufficient capacity" for the coursework. Those with better preparation, which often meant better family standing, would likely have these traits. But Rogers was sensitive to reaching beyond any single class of student. He actively searched for those with aptitude, taste, and capacity from the laboring and industrial classes as much as any other, establishing connections with the Massachusetts Charitable Mechanic's Association, an organization that supported the advancement of working-class youth in trade and technical work. Their discussions were aimed at supporting and enrolling a small number of students aided by the association. Whatever the social background, any student wishing to complete the requirements for the degree, according to Rogers, must have "zeal and energy." Rogers, often described by others as a man "incessantly busy" and of "irrepressible activity," didn't look favorably on indolence and encouraged faculty members to identify students who did not have the vitality to complete the program.[32]

Rogers also hoped that students would approach their studies with an appreciation of the Institute's useful arts mission. Given its commitment to both practical and theoretical studies, students had extensive opportunities to engage in topics of both kinds of knowledge. This mission, according to student Charles R. Crop, gave the Institute one of its most attractive qualities. Crop, later a member of MIT's alumni association, appreciated the efforts the institution had made in "fostering the highest scientific attainment and endeavors, and at the same time, in consequence of this, be a source from which the whole community may draw its most practical workers and teachers." Students drawn to both the applied and the abstract would find a curriculum aligned with their interests. When students didn't agree with or understand this mission, Rogers sometimes took it upon himself to meet with them individually. He kept a log of these meetings to assess student concerns and the state of the Institute's instruction. In one such case, when a handful of students sent an open letter to the MIT Corporation about being deprived of a "thorough, complete and systematic instruction" in mechanical engineering, Rogers responded within days. Upon interviewing them, he discovered concerns common to students of any era: that the textbook was "too difficult to understand"; that the professor, in this case Channing Whitaker, breezed through topics too rapidly and "attempted too many subjects"; and that classroom exercises were "not practical enough." Other students interviewed commented that complaints from this handful were "too sweeping and strongly expressed." From these meetings Rogers concluded that delays and problems with laboratory facilities had impeded thermodynamics instruction and had given the impression that the subject was unimportant. While understanding the students' criticism, he also recognized that one student, who represented a few of his peers, only partially understood the useful arts approach taken by the Institute: "They were impatient to get to what they sought—more practical work, although in the opinion of himself and others—this [theoretical] subject was a fundamental one." If these practical-minded students recognized the significance of fundamental principles of science, then they were one step closer to the useful arts ideal Rogers had in mind.[33]

As he kept an eye on enrollments and the perceptions of students, Rogers also followed one of the great and pressing student-related concerns facing institutions of the era: coeducation. "Scientific" studies appeared in the 1870s that tended to support long-held ideas about separate spheres, the notion that men (in the world of work) and women (in the place of the home) best operated in separate domains. During MIT's first decade, one in which the Institute's members were considering the question of coeducation, Harvard professor of medicine Edward Clarke offered what were then considered evolutionary-based theories on why women and men shouldn't receive the same education. The physical and physiological constitutions of each had evolved and differed to such an extent, he argued, that separate forms of education were essential for the proper development of both. While his specific scientific interpretations received a mixed reception, Clarke's theories and those of others who followed, such as psychologist G. Stanley Hall, reflected broader social and cultural values that implicitly resisted equal forms of education for both sexes.[34]

In this context the members of the MIT community asked themselves whether they would admit women into their regular program. Internal discussions among faculty and MIT's government revealed a wide range of views. Although as recent scholarship suggests, it was common for young women to receive science and mathematics education at the precollegiate level, some at MIT held firm opinions against the idea of full coeducation; others who wanted to diversify the student population treated the topic gingerly. Rogers had already made clear that the Institute's popular, evening scientific courses would be open to both sexes. He met little resistance in making the case for this part of the instructional program. If he chose to expand the idea to regular courses of instruction, that would be a different matter. In many ways he stood between two competing interests, for some MIT supporters, donors, and influential government members had little appetite for changing the composition of the student body; others on the faculty pressed for precisely such changes. Rogers leaned toward inclusivity, as he had with his antebellum reform proposals, but he also struck a delicate political balance to avoid alienating major sources of support.

When four women who had attended evening classes in the late 1860s approached the Institute about taking regular classes in chemistry, Nathanial Thayer, a major donor, caught wind of the development and inquired about what the institution planned to do. Rogers responded to both Thayer and the students by saying that full coeducation couldn't happen "without seriously embarrassing the organization of the laboratory and other departments of the school as connected with the regular courses now in progress" and closed off the option for the time being. With the Institute only a few years old at the time, Rogers dodged a challenge that was likely to complicate the school's standing in an era unaccustomed to coeducation. How he responded to Thayer and others suggests that he was torn between ensuring the economic foundation of MIT and his democratic ideas about science.[35]

Rogers's initial move against coeducation stood in contradiction to his views on women in science more broadly. At the Boston Society of Natural History discussions in the 1870s surrounded whether the admittance of women should be entertained. Should they be allowed into the meetings? Would there be topics discussed that were unsuitable to women? The society's president at the time, Thomas T. Bouve, was of the opinion that women were "not at all interested, or but slightly so, in science," as he brought forward the question of admitting women. Rogers, knowing women had shown strong interest in the Institute, advocated that the society should adopt a system of grading membership without considering the sex of applicants. Although the proposal met with resistance and ultimately came to nothing, it reflected his ideas about women, science, education, and professionalization. On this count Rogers saw no reason to exclude merely on the basis of sex.[36]

After the Institute had survived its first five years in operation, Rogers began to encourage greater involvement of women in the Institute's regular program. Of the women who attended the evening lectures, many continued to ask about the daytime laboratory experience offered only to men. Ellen Swallow was one of those who wanted such experience and applied for regular student status for the fall of 1870. At the time President Runkle knew that trying the coeducational experiment, with safeguards in place, wouldn't conflict with Rogers's plans, for he led the institution guided by the belief that Rogers had the "final word" on substantive decisions. Runkle admitted Swallow without tuition for the spring of 1871. "I thought the President of MIT," wrote Swallow of her acceptance, "remitted my fee out of the goodness of his heart, but later I learned that it was because he could say I was not a student, if anyone should raise a fuss about my presence in the laboratories." Her admittance nevertheless paved the way for additional female students. In 1876 the Women's Educational Association offered funds to MIT for the foundation of the Women's Laboratory. The laboratory aimed to meet the needs of MIT's female student population, which had increased since Swallow had been admitted, graduated, and become an instructor there. Once Runkle, Rogers, and the faculty agreed to the new laboratory, Swallow remarked that "the only separation of the sexes is in the lab work. I believe no other scientific school in the world can say as much."[37]

The *Atlantic Monthly* followed the developments in MIT's student body and interpreted the new laboratory as a way to accommodate larger numbers of students. "A guaranty," reported the magazine's education editor, "was added that in any laboratories which might be built for the Institute in the future, provisions should be made for advanced instruction without distinction of sex." The Woman's Laboratory therefore meant not only new equipment and a building, but also a revision in the Institute's policy on coeducation.[38]

Delighted by these developments, local philanthropist and women's rights supporter Marian Hovey contacted Rogers to see what she could do in support of the coeducational movement at MIT. The idea of the Women's Laboratory caught her imagination and led to her decision to offer the Institute a gift of ten thousand dollars. She explained to him that, given MIT's direction, she saw no need to word the offer "as to include women. . . . it seemed better not to use any exclusive phrase." The purpose of the donation, to her mind, was to give "appreciation of the great practical work which the Institute is doing as well as from admiration for the justice and common sense with which women students are treated." Rogers thus found a way to merge a broader, democratic vision of American science with local philanthropic support.[39]

He also knew that he must proceed carefully if the coeducational movement were to last. His cautiousness came to the fore when discussing coeducation with Edward Atkinson, a prominent Boston financier, original MIT incorporator, and longtime government member. Atkinson's voice was one that Rogers couldn't ignore, given his standing in the community and his impassioned support for the institution. While some viewed his voice as mere "meddling" in Institute affairs, Rogers tended to take it seriously. Such was the case when it came to the issue of coeducation.[40]

Atkinson and Rogers had differing views about when and where to permit female students. Following the Hovey donation, Rogers proposed expanding the rights of women who attended during regular classes. He noted to colleagues and government members that women at the Institute were engaged in activities beyond traditional student functions. Women also served as faculty assistants, conducting work in departments and in connection with the library. The question he put before them was one regarding the separation of sexes between laboratories. Although both have been admitted as regular students, "is there any special action needed for determining the status of women as regular students?" "Having been regularly admitted," he continued, "should they not have all the privileges of regular students?" He implied that the distinction between labs seemed untenable in the long run and that the charge of faculty was to teach the same material regardless of place and student population.[41]

Atkinson disagreed. While he had little problem with including female student names in the catalogue he as regular students, he didn't think their inclusion as regular students should mean that they also receive all of the same privileges. "I think it is not expedient that the women should go into the regular laboratory," he explained, "[for] we have provided fitly for them." By "fitly," he was referring to the Women's Laboratory and, most likely, supplies and support generated from the Hovey donation. If Rogers saw the distinction as artificial, however, he still went along with it. Exactly why he followed this direction and what purpose he had for doing so isn't clear. The common concerns—maintaining economic security and pleasing influential supporters—likely stood at the center of his decision to keep the status quo. "The young women entering as regular students in the chemical courses," he informed faculty, "shall be expected to pursue their chemical work in the women's lab under the direction of the Professors in charge of the same, who shall arrange their studies and examinations and judge their proficiency." But in the end Rogers won only a partial victory. On the one hand, he had witnessed the regular student body change from single-sex to coeducation and an expansion of laboratory facilities. On the other hand, the rights of regular female students to work side by side with men in their chemical work still stood in the distance.[42]

❧

Whatever their background or gender, all students began to see developments in the curriculum that raised questions about the purpose and mission of the Institute. Students who enrolled at MIT for the first two years followed a prescribed curriculum in general preparation for advanced scientific study. The first and second years focused on mathematics, mechanical drawing, freehand drawing, elementary and experimental mechanics, chemistry, English language and literature, and modern languages. The second year included extra work in navigation and nautical astronomy as well as surveying. "Up to the end of the second year," read the Institute's first catalogue, "the studies are the same for all regular students; each thus obtaining such an acquaintance with the whole field of practical science as is needed for the further pursuit of the studies of the School, in any of its departments." In the third and fourth years, however, students could select specialized course work for one of six majors: mechanical engineering, civil and topographical engineering, practical chemistry, geology and mining, building and architecture, and general science and literature. With the exception of mechanical engineering, the curriculum included extensive laboratory practice.[43]

During Rogers's leave of absence, Runkle wrote that mechanical engineering students needed a substantive laboratory in which to learn. As early as 1869, Runkle began to make arrangements for students in this area to gain field experiences off-campus. At first he arranged for them to work in machine shops at the Navy Yard in Boston. In years that followed he offered students professional field trips, but all the while Runkle continued to search for a system of laboratory instruction to match the Institute's theoretical and practical aims. Acknowledging that the Worcester Institute, Cornell University, and Illinois University each had "built up shops, but always from the manufacturing side and idea, and not from the teaching side," he explained to Rogers that MIT needed something similar.[44]

In June 1876 Runkle believed he had found the answer to his problem at the Centennial Exhibition held in Philadelphia. Of all the displays at the science fair, the Russian exhibit made the most significant impression on him. "In an instant," he explained, "the problem I had been seeking to solve was clear to my mind; a plain distinction between a Mechanic Art and its application in some special trade became apparent." The "plain distinction" he referred to had to do with separating the skills needed in a trade from the skills to be taught at the Institute. In industry various trades employed similar mechanical skills and tools in a variety of sequences for such unrelated purposes as constructing machines for a mill factory and building a stream engine. Rather than teach each mechanical trade, however, the Russian system had constructed a model of training that would provide laboratory experiences for skills required across multiple trades without necessary reference to any particular application. Ecstatic over his finding, Runkle remarked that "making the art and not the trade fundamental, and then teaching the art by purely educational methods is the Russian system." The Russian model had shown him that "the arts are few, and the trades many." He immediately undertook an effort to develop the system at MIT and, in doing so, hoped to popularize the method in the United States.[45]

Runkle first turned to Rogers for support. He told the him that "Russia has taught us a grand lesson. You know that the workshop problem as part of the course for Mechanical Engineers has been a difficult one." The laboratory experience, he argued, had provided satisfactory results in the other fields studied at the Institute. With a similar offering, "our mechanical engineers . . . would be independent when they graduate, instead of being, as now, the most helpless product of any of our departments." He explained that the work conducted by students in the system emphasized developing knowledge and skills rather than constructing products of marketable value. Runkle projected that students would learn exemplary uses of tools, materials, and designs as opposed to the manufacture of objects. Rogers and the MIT government found the arguments compelling and agreed to implement the program in 1876.[46]

MIT's new program, the School of Mechanic Arts, received considerable publicity. In the wake of such attention Runkle's plan gained its share of supporters and critics. Supporters cited the efficiency with which the new system would impart skills. Graduates of the program, reported one review, "may not be first-rate journeymen carpenters and machinists, but they are advanced beginners, and have a better general idea of the theory and practice of their trade than the average workman in it." Another reviewer wondered how the Institute would avoid offering technical education that "is in the direction of a too long, minute, and elaborate preparation." The Russian system, it was hoped, would be a more efficient means of instruction in this area. "Precious time is lost before theory is aided by practical illustrations and action. This evil of overdoing the preparation for technical instruction is to be avoided by the workshops of the Institute of Technology."[47] Still others looked optimistically to the new program as a grand step forward beyond the apprenticeship system. As one reviewer put it: "Herein lies the marked difference between education which the student receives in such a school and that which he receives in a shop—a machinist's shop, for instance. There, once he has learned to do a thing well, he is kept at work upon it, because his labor is useful to his employer; here, once he has learned a process, he is advanced another degree, because his education, and not his availability, is the primary consideration."[48] Runkle's supporters believed that the plan attempted to satisfy two purposes: efficiently preparing students seeking advanced scientific instruction and of those entering industry after two years of study in the Mechanic Arts school.

Critics were less optimistic about the Russian system. One believed that the MIT school made a start at "a movement in the right direction" but predicted it wouldn't ultimately succeed. A similar program initiated at Girard College had "utterly failed" and resembled attempts at MIT to carry out the same enterprise. Its success would largely depend on if it received the proper "support" and "sympathy." The advocates for "real experiences" found that the mechanic arts workshops failed to prepare students for on-the-job realities.[49]

Runkle's system, in the end, met neither of the two basic goals it had proposed to address. Students seeking advanced scientific training were uninterested in what the School of Mechanic Arts had to offer. Those seeking basic skills in the two-year program found the training insufficient for anything above entry-level positions in industry. After slightly more than a decade in operation, the MIT Corporation dissolved the Mechanic Arts school.

The freedom to experiment with the curriculum was an important factor that distinguished MIT from established institutions such as Harvard. The Institute, in large measure, stood for this kind of autonomy to start and discontinue programs without interference. The opening and closing of the Mechanic Arts school was part of the process that MIT promoted in trying alternative forms of instruction to meet the needs of abstract and applied sciences. To critics Runkle's failed experiment with the School of Mechanic Arts meant that the idea of MIT was not yet firmly established. For supporters the expansion of enrollments at the Institute and the perceived decline of Harvard's Lawrence School provided evidence favorable to the Institute's standing. More important, the efforts of Charles W. Eliot to alter MIT's relationship with Harvard provided a conspicuous measure of the Institute's reception for both critics and supporters.[50]

When Eliot resigned from MIT in July 1869, he believed the Institute had established a niche in American higher education by experimenting with instruction and the curriculum. Rogers hoped to keep Eliot on the faculty at MIT, but the chemist had been offered Harvard's presidency. With Runkle in command at MIT, Eliot looked to Harvard to satisfy his administrative interests. "It will be a loss to us," mentioned Runkle to Rogers, "but it will also be a gain to have a President at Harvard who believes that the mission of the two institutions is distinct, and that there should be no jealousy or rivalry between them." At first neither Runkle nor Rogers suspected the merger plans then simmering. Upon leaving for Cambridge, Eliot praised Rogers's "example and precepts, . . . wisdom and wide experience." As for relations between MIT and Harvard, Eliot commented, "I mean to see that the Institute enjoys the field it has so honorably won, without competitions or duplication of any sort at Cambridge." With these assurances the chemistry professor was installed as Harvard's president in the fall of 1869. Soon after, however, Eliot began his effort to unite the two institutions. He gave every indication of his interest in the type of training offered by the Institute, especially in his inaugural address. Harvard should have "science taught in a rational way, objects and instruments in hand—not from books merely, not through the memory chiefly, but by the seeing eye and the informing fingers." Much in the spirit of Rogers's idea for MIT, Eliot emphasized that "the actual problem to be solved is not what to teach, but how to teach." Even with such allusions, many Institute members missed signs of the coming controversy.[51]

By January 1870 Runkle sent Rogers word of merger schemes brewing at Harvard. Runkle described Eliot as "full of the idea of consolidation of the Institute and Harvard University" and warned Rogers to expect a proposal from Harvard about a merger. Because it wasn't the first time Rogers faced such a proposal, he had a ready answer. "I can see nothing but injury to the Institute," he told Runkle, "from the projected change. . . . Those who know our History know that [MIT's] success is due to the Opportunity we have had under the inspiration of Modern ideas. No kind of co-operation can be admitted by the Institute which trenches in the least degree upon its independence." Rogers clearly had no interest in having Harvard absorb the Institute.[52]

Regardless, Eliot continued to press for a merger. From Rogers's point of view Harvard envied "the Institute [which] has already taken the first place among the Scientific Schools of the U.S." With the laboratory method squarely at the center of the Institute, MIT had attracted attention for its experimental scientific instruction. For Eliot, Harvard's Lawrence School seemed in a state of decline, failing to attract students or prestige. When compared to the Institute, one scholar described Harvard's program as a "dead carcass." Eliot himself would later bemoan the shabby state of science at his institution. "The reason why the School is dying is simply this," he wrote to a benefactor, "the Sheffield School at Yale and the Institute of Technology at Boston have many more teachers, a better equipment, and a vastly greater variety of instruction." Not surprisingly, the Harvard president felt pressure to do something effective with the large Lawrence legacy. His scheme was to transfer the Lawrence funds as well as those of the Bussey bequest to MIT, in return for making the Institute a "Faculty" within Harvard, similar to the Harvard Medical faculty in Boston. Eliot mentioned the plan to Runkle, assuring him of MIT's independence to run the school as free from Harvard control as the Medical School, a "Completely Independent body." As for retaining the character of its founder, Eliot suggested, "it would be right to give the combined institution the name of 'Rogers,'" as in, perhaps, the Rogers School of Science. The composition of the governing bodies of Harvard and MIT, moreover, suggested that the plan might not face much resistance. Three from the Harvard Corporation were also original members of the Institute, and nine of forty government members from the Institute were graduates of Harvard.[53]

Convinced of the merits of his scheme and the inevitability of its implementation, Eliot began making arrangements to visit Rogers personally in Philadelphia. Runkle, himself a graduate of the Lawrence School, didn't actively rebuff Eliot's overtures. But he warned Harvard's president that "the Inst. is simply what Professor Rogers has made it" and that "no one . . . will hesitate to accept any opinion which he may have in the matter as final." Rogers still suffered from his "nervous exhaustion" at the time but decided to make his position on the merger perfectly clear. Interrupting Eliot's plan to visit Philadelphia, Rogers assured him that "no contribution of funds would justify us in consenting to change." Rogers had no doubt that a merger between Harvard and MIT "would be a decided disadvantage to the Inst., which owes its success in great measure to the fact that it has stood entirely uncon-nected with other institutions." Without hesitation he placed autonomy before all else.[54]

Eliot didn't give up. He simply began to talk independence with Rogers, about how the merger would not make MIT any less independent, about how the Insti-tute would retain its property and organization. He replied to Rogers that MIT "would be stronger and more independent than now" if it became "an independent department of . . . [Harvard] University." Rogers dismissed Eliot's argument that the merger would preserve the Institute's autonomy. Runkle and a majority at MIT sided with Rogers, while Eliot positioned the Harvard Board behind the scheme.[55]

Much of this activity occurred in the few months between Rogers's resignation from the Institute and Runkle's official election to the presidency. Eliot saw this pe-riod as a window of opportunity to maneuver his plan into action. Almost the in-stant he assumed the MIT presidency, Runkle agreed to committee meetings with Eliot, each president bringing two officers or faculty from their respective institu-tions. When Runkle reviewed some of Harvard's generous offers and detailed plan for autonomy, he asked Rogers, "Shall we take it?"[56]

Runkle wrote with more enthusiasm for merger than ever following these meet-ings during the fall of 1870. He knew Rogers would disagree with a provision in the latest merger plan that required all changes to courses, departments, instruction, ad-mission, and graduation to be approved by Harvard. Runkle acknowledged that "this would not leave us much after all." But he also argued that "if we do not unite, & do not get the means . . . to raise salaries we shall lose all the Professors we have whom we could least afford to spare." Rogers brushed aside the arguments and, by helping to defeat Eliot's plans, left little doubt about his desire for MIT to continue as an independent center of science instruction and research.[57]

❦

The general reception of the Institute, as measured in public discourse as well as the school's enrollments, curriculum, and relation to Harvard, would continue to preoccupy Rogers. But for a time he turned to matters having to do with science and professional service. As always, he approached work in these areas with both the practical and theoretical in mind—a conviction around which he had organized his idea of the Institute and one he had often presented to the public.

During his ill health between 1868 and the mid-1870s, Rogers revisited his inter-ests in geological research and the professionalization of science. He had never felt finished with his survey work in Virginia after the state declined to publish his final report. Rogers's years of notetaking, mapmaking, and laboratory work had not yielded a full and complete public record. During the Civil War the Union army commissioned Thomas Ridgeway, a former assistant on Rogers's survey, to construct a map of Virginia. Ridgeway produced a map based on notes collected on the geological survey. As the war map fell into obscurity after Appomattox, Rogers planned to prepare a final geological map and report for the state. To this end Jed Hotchkiss became Rogers's best hope for completing the project. Hotchkiss, a Virginia educator and free-lance engineer, had gained surveying and mapmaking experience as a Confederate soldier. In 1873 Hotchkiss's work with the Washington College Board of Surveying led him to collaborate with Rogers. The result was published three years later in *Virginia. A Geographical and Political Summary.* Set to a scale of twenty four miles to one inch, the map brought Rogers closer to his ambition. Yet in subsequent years Rogers and Hotchkiss planned a more detailed map, one with a scale of eight miles to one inch. Because of health reasons, Rogers couldn't contribute to the completion of a larger map.[58]

In 1875, and over the next five years, Rogers managed to publish works on Virginia's deposits, maps, geological formations, and iron ores. In addition, he wrote on the notebook records gathered from field researchers on the survey. Despite these efforts in the twilight of his career, he'd only scratched the surface of producing a thorough survey report, which he had long desired to complete.[59]

As for professionalization, Rogers focused his efforts on the National Academy of Sciences (NAS) as it continued to serve its function of providing scientific advice to government agencies. Predictably, he fell into conflict with old guard Lazzaroni members at the academy. By controlling of the membership list, some of these members continued to oppose Rogers and the faction he represented. Shortly before the onset of his illness in 1868, he'd failed to attend some of the annual meetings, causing his name to be "stricken from the roll of members." He suspected foul play when the NAS declared "non-attendance . . . without communicating to the Academy a valid reason for his absence" as the reason for revoking his membership. Upon receiving notice in 1872 of his reelection to the academy, Rogers decided to contact NAS president Joseph Henry directly about a similar matter concerning his brother Robert. William refused the offer of reinstatement if his brother Robert remained a nonmember. "As he became a member of the first organization with me," explained William, "and was dropped from the list at the same time and for the same cause . . . I could not consent to be the object of any partiality in the matter of reappointment." With his notice to Henry, William issued his resignation to the academy. Joseph Henry accommodated the request and ultimately installed both brothers as members once again.[60]

After his reinstatement in the National Academy of Sciences, William's activities within the organization steadily grew. During his absence the academy had suffered setbacks, including a dwindling membership and a complete halt to government requests for advice. For a time in 1870 some scientists sought to disband the organization. Astronomer Simon Newcomb thought it "increasingly doubtful whether the organization would not be abandoned." By the time Rogers returned to the NAS, then president Joseph Henry had managed the problems by expanding membership rolls, reducing to one the number of meetings in a year and requiring members to have an established program of published research. With such moves Henry kept together the academy as an honorary society. The first substantial work offered to the NAS came in 1878, when Acting President O. C. Marsh elected Rogers and five other scientists to help resolve survey disputes in the western territories. The project breathed new life into the organization and gave Rogers an opportunity to exercise his geological experience.[61]

Rogers and the committee began a review of five major surveys of western territories that had been commissioned simultaneously in the 1870s. Wrangling and disputes emerged between the expedition parties involved, which included Union Pacific and Central Pacific railroads, the Corps of Engineers, the Department of Interior, the Smithsonian, and the Treasury Department. As their interests clashed, Congress requested the academy to consider potential resolutions to the survey disputes. The committee reported to Congress on plans, methods, and expenses for a survey of the western territories. The recommendations gave direct rise to the U.S. Geological Survey housed by the Department of Interior. The survey also became the first large-scale agency wholly conceived by the academy and acted upon by the federal government. Marsh commended Rogers for his efforts on the report to Congress, noting that it was "as well received in Washington as it was by the Academy."[62]

Developments with the survey in part paved the way toward the selection of Rogers as the third NAS president, following the presidencies of Bache and Henry. On the day of the election in April 1879, the academy informed Rogers by telegraph from Washington of his election. Surprising attendees such as Harvard's George J. Brush, who imagined the nearly seventy-five-year-old president in frail health, Rogers arrived the next day to deliver an acceptance address. As a result of health problems, however, Rogers served as the academy's president for only three years, half the normal term for the office. During his tenure three basic developments occurred. First of all, a recently formed National Board of Health requested the NAS assist in formulating a plan to organize a nationwide public health program. The board had been created to respond to a yellow fever epidemic spreading along the

Mississippi River in 1878. Their request of the academy resulted in a collaborative effort between the two organizations in issuing a report submitted to Congress in 1880. They recommended a plan for organizing both military and civilian health workers to respond to epidemics of the kind witnessed two years earlier. The second development came out of increasing concern over the condition of the original Declaration of Independence. Congress sought advice from the NAS on what to do with the aging, fading document. After determining part of the cause of deterioration—such as the fading of ink, the wear caused by "press copies," exposure to the elements—the academy ran tests and determined that chemicals wouldn't preserve the document. Instead, Congress abided by the NAS's suggestion to secure the document behind a wooden enclosure. The third request came from Rogers, who initiated a movement to restart publication of the organization's proceedings. He believed that the papers delivered by NAS members at their annual meeting would better serve the advancement and diffusion of science if supported by an annual publication. Out of the movement came the NAS *Memoirs*, a publication that ran until the opening of World War II.[63]

Despite his age and flagging health, Rogers received offers to hold other offices during and after his leave of absence from the Institute. The American Association for the Advancement of Science elected him president in the 1870s. Although scheduled to present his research before the AAAS, he had to decline for health reasons. When elected president for the Boston Society of Natural History a few years later, he flatly declined the offer for health reasons as well. The same issue caused him to step down from offices in local Boston societies, such as the Thursday Evening Club and the Saturday Club. By the fall of 1874, however, Rogers had returned to the area and was entertaining club members at his home with Emma.[64]

What these research and professional appointments suggest is that, whenever possible, Rogers continued to present a public persona for the useful arts ideal. His commitment to the geological studies of Virginia kept him engaged with basic scholarly questions in earth science as well as with practical efforts to develop knowledge of southern natural resources. The professional appointments gave him a platform from which to advocate the interests of both practitioners and theoreticians. He thus managed to turn public perceptions of the Institute from what they had been after the inaugural addresses of Bigelow and Atkinson to what he had envisioned since his first efforts at higher educational reform. By this point in Rogers's remaining years, he and the Institute had become one and the same in the public eye. In the words of John Runkle, who at the time led the Institute as president, MIT "is simply what Rogers has made it."[65]

When Runkle stepped down from the presidency of the Institute in 1878, it's not surprising that many looked to Rogers to assume the post. Yet the empty coffers at the Institute, not to mention the founder's physical condition, made a return unlikely. He made clear that the only way he'd take the job would be if the MIT government made a commitment to raise $100,000 to renew the dwindling endowment and begin an immediate search for a successor. They called his bluff.[66]

Before the Institute had time to settle into a second Rogers presidency, he began conferring with the school about having Francis Amasa Walker as his successor. Walker had achieved distinction as a social scientist at Yale and as the director of the 1880 U.S. Census during President James A. Garfield's administration. After discussing the opening with Walker, Rogers received authorization from the MIT Corporation to deliver a formal offer. "I now write," he stated in June 1880, "to offer the position to you, and I need not say my dear Prof. how earnestly I desire that you will accept it." Delays in Washington prolonged Rogers's stay at the Institute until November 1881, when he had the opportunity to introduce Walker to the faculty formally. Walker complemented the science faculty with his offerings in political economy, a match that would help broaden the technical training of MIT students. From that point forward Rogers felt "released from the cares of MIT" and free to pursue his unfinished projects.[67]

CHAPTER NINE

This Fatal Year

ALTHOUGH HE'D RETIRED from his duties, Rogers gave no indication that he planned to stay away from MIT for very long. Within weeks of handing the presidency to Francis A. Walker, he accepted an invitation to deliver the Institute's May 1882 graduation speech. On that fateful day Rogers, in a very literal sense, gave his life to MIT. According to the physicians who rushed to the podium when he fell, "Life was extinct before his body fairly touched the floor." When news of his death reached his friends and colleagues, their responses came pouring in. Letters from former students, fellow researchers, educational leaders, politicians, philanthropists, and many others expressed their sympathy and loss to Emma. They warmly shared similar views of the man—the way he inspired his students to love science and the "sense of obligation" they felt toward him; his ability to touch others with "wonderful power of illustration and expression"; his personal dedication to "all that concerns the progress of science"; the "combined feelings of affection and respect" for him as a person and a scholar. "How few of us are left," despaired Asa Gray, "after the mortality of this fatal year."[1]

The year was "fatal" because Darwin had also died just a few weeks earlier. Their nearly simultaneous deaths were momentous to Gray's generation—they signaled the end of an era in science.

In terms of scientific research Rogers was, of course, no Darwin. He had contributed no grand theory of evolution, field-changing insight in geology, or revolutionary idea for natural philosophers. Rogers was, rather, a middling scientist among his peers, a determined, creative scholar with a penchant for understanding how different kinds and levels of scientific knowledge could inform one another. His determination led him to publish over one hundred works examining vexing geological and natural philosophical problems. His efforts, as with the Virginia Geological Survey, generated a great wealth of data. This Baconian enterprise laid the foundation for his more creative scientific work developing theories about mountain chain formation. More important, however, is what Gray undoubtedly had in mind when he thought of that fatal year. Darwins of any era rely on the labors of scientists like Rogers for their own work. Before going public with *Origin,* Darwin turned to places like Gray's Botanic Gardens at Harvard, that great repository of plant specimens from all over the world, for supporting data. Without the data-collecting work of Rogers, Gray, and like-minded researchers, grand theories amounted to little. Thus, Darwin stood in part on the shoulders of middling scientists, and Asa Gray understood this as well as anyone.

Rogers's death robbed Gray of a close colleague who shared in defending Darwin's theory, in collecting data necessary for theorizing, and in attempting to keep scientific institutions out of the hands of the exclusive, sometimes secretive Lazzaroni. They had taken on these struggles together. They had lived to see science mature and loosen itself from the hold of the past, represented by figures such as Agassiz. With Rogers, Darwin, Agassiz, Joseph Henry, A. D. Bache, Benjamin Peirce, and many others gone, that chapter in American science was drawing to a close.

Throughout his long scientific career Rogers followed many goals and interests, but if there was one legacy he hoped to leave through his research, it was advancing the interdependence of theoretical and practical questions in science. For most of his life the scientific community had been divided. Advocates of basic science, like Agassiz, had little interest in the work of practical scientists such as Jacob Bigelow. Rogers's articles, books, and presentations stood out in his day as work conducted by someone who defied boundaries and restrictions others sought to maintain. All branches of scientific inquiry, he believed, needed thorough, sustained investigation. As if in a nod to this view, the American Association for the Advancement of Science (AAAS), a few months before Rogers gave his final speech, changed its organizational structure. The AAAS divided the physical sciences division in two, one section for theory (i.e., physics) and the other for practice (i.e., engineering). They carved out a space for specialists in each area.

Were it not for Rogers and his passion for and approach to scientific research, there would be no MIT today. The conclusions he reached while collecting specimens in the field or experimenting with materials in the laboratory inspired Rogers's ideas about higher education reform. He wanted an institution that would train the next generation of scientists who took seriously the interplay between different levels and forms of scientific investigation. MIT's mission, he made clear, stood for the commingling of theory and practice. Rogers believed that the European emphasis on theory, on the one hand, had a greater role to play in the American science. On the other hand, he argued that America's fervor over technology provided better, more accurate tools with which to improve scientific theories. The Institute he envisioned brought the two traditions together in a laboratory-centered system of higher learning. At the commencement of 1882 Rogers told students that their education had equipped them for "practical industries" as well as for research in "the laboratory or in the field." The "thoroughness" and "accuracy" of student work at the Institute reflected the useful arts ideal.[2]

Not all scholars shared this vision. A younger generation of scientists who advocated either pure or applied science was critical of Rogers. The Johns Hopkins University physicist Henry A. Rowland complained of "professors who degrade their chairs by the pursuit of applied science instead of pure science." Faculty who squander their "energy and ability in the commercial applications" of science, he warned, represented a "disgrace both to him and his college." Citing the lack of pure studies, when compared with Europe, Rowland employed nationalistic arguments in calling for a refocusing of priorities in American science. Other researchers, such as engineer Robert H. Thurston, argued for an emphasis on practice over theory. Through his work at the Stevens Institute of Technology and Cornell University, Thurston defended utilitarian studies, particularly in such areas as mechanical engineering, and dismissed the need for theoretical or abstract studies. In these programs he discouraged his students from abstract mathematics and related coursework for fear it would distract them from applied science.[3]

Rogers believed MIT could bridge this divide through laboratory instruction, and in many ways it did. While the Institute alienated some scholars of the late nineteenth century, others gravitated toward MIT's catholic approach. It offered a vision of laboratory work that became one more piece of a broader applied studies movement encouraged by the federal land-grant legislation and by the emergence of the modern university. After a brief tenure at MIT, Charles W. Eliot took the laboratory ideal to Harvard, where his leadership helped shape a truly national university. At the start of his presidency in Cambridge, Eliot reminisced about Rogers's "example," confessing, "I received from [his] School much more than I ever gave." Once installed in Harvard's bully pulpit, Eliot preached the gospel of the laboratory. "The old-fashioned method of teaching science by means of illustrated books and demonstrative lecture," he assured college leaders, "has been superseded . . . by the laboratory method, in which each pupil . . . works with his own hands, and is taught by his own senses." The spark of reform leaped to the desk of Princeton's president John McCosh, who wrote in 1877 that "there is a growing feeling that scientists cannot be trained by mere lectures." During his administration McCosh led a drive to provide students with laboratory instruction beyond the lecture demonstration. Similar ambitions came alive at Amherst, where facilities for laboratories began to appear. In November 1876 a student observed that "Professor [Elihu] Root had introduced a novelty to the department of physics. Lab work is to be performed in connection with study. For this purpose the room in Walker Hall, known heretofore as the Alumni Room, is to be used." Yale had long before tolerated laboratories at its Sheffield School, but undergraduates at the college complained, through most of the nineteenth century, that "instruction was given, in large measure, by lectures, and these were not accompanied by strict requirements of personal investigations on the students' part." When the university revolution came to New Haven, faculty began transplanting the practices of the scientific school to the college proper.[4]

MIT's presence on the collegiate landscape prompted questions about the need for change. The Institute's John Runkle observed that other schools were following MIT in "moving in the matter [of laboratory instruction]": "I heard privately that [Harvard's] Prof. Gibbs intends to attempt something of the kind soon." Agassiz, Gibb's colleague, decried "the imminent danger in which our University is of losing its prestige if rigorous steps are not taken to strengthen it in the direction demanded by the wants of the nation." Similar pressures appeared on the West Coast. John LeConte, of the University of California, Berkeley, asked Rogers about MIT's laboratory method, for, he explained, "your experiences [are] valuable to us": "Any document having reference to the programme of organization; to the internal arrangements of the laboratory; to its practical working, etc. would be acceptable. In short, anything which would assist me in the organization of such a department in the most efficient manner. Perhaps, more recent experience may enable you to add some valuable suggestions. I hardly think we shall be able to accomplish anything before 2 years from this time; but I wish to have my plans matured before hand." LeConte's inquiry, and those of others, reveals a spirit of curiosity and enthusiasm for change that was shared by many members of the scientific community. Thus, by means of its "example" the idea of MIT became part of the broader discourse in American higher education.[5]

Rogers's death not only marked the end of a scientific era but also highlighted the beginning of a new educational outlook. MIT popularized a model of laboratory instruction that was absorbed elsewhere (although controversy never wandered far from this model, as evidenced in the later *Atlantic Monthly* debates over MIT between Francis A. Walker and Harvard's Nathanial Southgate Shaler).[6] Walker's interpretation of the MIT ideal, as originally defined by Rogers, remained largely unchanged throughout the closing years of the century. For the most part the institution continued to focus on the needs of the pure and applied science community. Except for minor changes to the Institute's governance and curriculum, such as the introduction of a physical education program, it was the original idea of MIT that Walker promoted to philanthropists and to the state legislature. During his tenure this approach persuaded Massachusetts legislators to grant the Institute $300,000 for expanding its facilities and establishing scholarships for qualified state residents who were unable to pay for tuition. Such developments enhanced competition for students, resources, and prestige, prompting Harvard officials to renew their call to merge with MIT in the 1890s and then again in the 1910s. While the takeover plans proved unsuccessful, the ideas behind both institutions established factions in higher learning that continued to be the subject of public interest into the twentieth century.[7]

William Barton Rogers left behind ideas about higher education that college leaders would continue to debate, and his wife, Emma, helped secure his legacy through her interest in his research and in the Massachusetts Institute of Technology. Following his death, and after fielding "frequent requests . . . by geologists and others" for pieces of his work, she decided to collect and republish her husband's research. In 1884, two years after his funeral, Emma finished editing a volume of approximately eight hundred pages of Rogers's publications, reports, and maps. The compilation included his early work on marl, his annual reports on the Virginia survey, and an assortment of papers that described or generalized about geological formations in the South. Emma also made thoughtful contributions to MIT's department of geology. She donated books, photographs, and funds for periodicals and microscopes to the department's library, and, when she passed away, in 1911, she left a substantial portion of the Rogers estate to MIT.[8]

In the end Rogers's career and ideas intersected with the principal values defined by the useful arts. He remained convinced that, whether it was in science, professionalization, or higher learning, theory or practice alone would not do. Throughout his career he advocated the view that both had to coexist and flourish before substantive gains could be derived from science and for society. Although his epitaph at Cambridge's Mount Auburn Cemetery reads simply, "William Barton Rogers, 1804–1882," the MIT motto, MENS ET MANUS (mind and hand), records the work of a lifetime.

Abbreviations

BSNH	Boston Society of Natural History
CWE	Charles W. Eliot
ES	Emma Savage/Emma Rogers
GV	William Barton Rogers, *A Reprint of Annual Reports and Other Papers on the Geology of the Virginias* (New York: D. Appleton, 1884)
HDR	Henry Darwin Rogers
JBR	James Blythe Rogers
JDR	John Daniel Runkle
LL	Emma Rogers, ed., *Life and Letters of William Barton Rogers,* 2 vols. (Boston: Houghton Mifflin, 1898)
PKR	Patrick Kerr Rogers
RER	Robert Empie Rogers
UVA	University of Virginia
WBR	William Barton Rogers
WBRP-MITA	William Barton Rogers Papers, Massachusetts Institute of Technology Archives, Cambridge, MA

Preface

1. MIT Society of Arts, *In Memory of William Barton Rogers, LL.D., Late President of the Society* (Boston: Society of Arts, 1882), 20–21.

2. Richard Westfall, *Never at Rest: A Biography of Isaac Newton* (Cambridge: Cambridge University Press, 1980)

3. MIT Society of Arts, *In Memory of William Barton Rogers,* 20.

4. Louis Agassiz to Governor John A. Andrew, December 16, 1862, John Andrew Papers, Massachusetts Historical Society.

5. Bruce Sinclair, "The Promise of the Future: Technical Education," in George H. Daniels, ed., *Nineteenth-Century American Science: A Reappraisal* (Evanston: Northwestern University Press, 1972), 249.

6. Emma Rogers, ed., *Life and Letters of William Barton Rogers,* 2 vols. (Boston: Houghton Mifflin, 1898). Archival documents suggest that Emma Rogers continued to use her maiden name; I refer to her as Emma Savage in this study.

Chapter One · An Uncertain Future

1. WBR was in Williamsburg, Va., from 1819 to 1825. During this time he enrolled as a student and assisted his father, who was William and Mary's professor of chemistry and natural philosopher. It is unclear, however, whether William completed all the requirements for graduation. The institution listed him as "in attendance" until 1821 (*A Provisional List of Alumni, Grammar School Students, Members of the Faculty and Members of the Board of Visitors of the College of William and Mary in Virginia from 1693 to 1888* [Richmond, Va.: Division of Purchase and Printing, 1941], 35). But the records for those who received degrees at the college during Rogers's years of attendance have been destroyed. Valuable manuscript collections on Rogers's early years as well as later family developments can be found in the Massachusetts Institute of Technology Archives holdings among the Rogers Family Papers; William Barton Rogers Papers; and William Barton Rogers II Papers.

2. PKR to WBR, October 17, 1825, *LL,* 1:31–32; WBR to PKR, November 3, 1826, *LL,* 1:34–36; WBR to PKR, November 3, 1826, *LL,* 1:36; *Picture of Baltimore, Containing a Description of All Objects of Interest in the City and Embellished with Views of the Principal Public Buildings* (Baltimore: F. Lucas Jr., 1832), 196.

3. WBR to PKR, January 25, 1827 and March 14, 1827, *LL,* 1:37–39, 40–41.

4. WBR to PKR, February 19, 1828, *LL,* 1:47.

5. WBR to PKR, January 25, 1827, *LL,* 1:37–38.

6. On the discrepancy between scientific knowledge and scientific offerings in American higher learning, see Mary Jo Nye, *Before Big Science: Pursuit of Modern Chemistry and Physics* (Cambridge: Harvard University Press, 1999); and Stanley M. Guralnick, *Science and the Antebellum American College* (Philadelphia: American Philosophical Society, 1975). Guralnick argues that scientific instruction in American colleges of the 1820s and 1830s continued to use long outdated texts and discarded ideas (62, 79). General histories on the state of science and professionalism in the United States include George H. Daniels, *American Science in the Age of Jackson* (New York: Columbia University Press, 1968); John C. Greene, *American Science in the Age of Jefferson* (Ames: Iowa State University Press, 1984); Robert V. Bruce, *The Launching of American Science, 1846–1876* (New York: Knopf, 1987); Sally Kohlstedt, *The Formation of the American Scientific Community: The American Association for the Advancement of Science, 1846–1860* (Urbana: University of Illinois Press, 1976); George H. Daniels, "The Process of Professionalization in American Science: The Emergent Period, 1820–1860," in Nathan Reingold, ed., *Science in America since 1820* (New York: Science History Publications, 1976), 62–78; Nathan Reingold, "Definitions and Speculations: The Professionalization of Science in America in the Nineteenth Century," in Alexandra Oleson and Sanborn C. Brown, *The Pursuit of Knowledge in the Early American Republic: American Scientific and Learned Societies from Colonial Times to the Civil War* (Baltimore: Johns Hopkins University Press, 1976), 33–69. On the American appetite for technological innovation and its relationship to science, see Edwin Layton, "Mirror-Image Twins: The Communities of Science and Technology in 19th

Century America," *Technology and Culture* 12 (January 1971): 562–80; Howard P. Segal, *Technological Utopianism in American Culture* (Chicago: University of Chicago Press, 1985).

7. Ruschenberger, *Sketch,* 1–4.

8. Preface, *American Philosophical Society, Transactions,* 1 (1771): xvii, cited in Greene, *American Science,* 6.

9. Edgar Fahs Smith, "James Woodhouse," *Dictionary of American Biography,* 22 vols. (New York: Charles Scribner's Sons, 1928–58), 20:491; Harriet W. Warner, ed., *Autobiography of Charles Caldwell, M.D.* (Philadelphia, 1855), 173, cited in Chandos Michael Brown, *Benjamin Silliman: A Life in the Young Republic* (Princeton: Princeton University Press, 1989), 104.

10. George Blumer, "Benjamin Smith Barton," *Dictionary of American Biography,* 2:17–18; David Y. Cooper and Marshall A. Ledger, *Innovation and Tradition at the University of Pennsylvania School of Medicine: An Anecdotal Journey* (Philadelphia: University of Pennsylvania Press, 1990), 18.

11. Richard H. Shryock, "Benjamin Rush," *Dictionary of American Biography,* 16:228; Donald J. D'Elia, *Benjamin Rush: Philosopher of the American Revolution* (Philadelphia: American Philosophical Society, 1974); David Freeman Hawke, *Benjamin Rush: Revolutionary Gadfly* (Indianapolis: Bobbs-Merrill, 1971); Carl Alfred Lanning Binger, *Revolutionary Doctor: Benjamin Rush, 1746–1813* (New York: W. W. Norton, 1966); Richard Hofstadter and Wilson Smith, eds., *American Higher Education: A Documentary History* (Chicago: University of Chicago Press, 1961), 153.

12. Ruschenberger, *Sketch,* 6–7.

13. Leonard G. Wilson, ed., *Benjamin Silliman and His Circle: Studies on the Influence of Benjamin Silliman on Science in America: Prepared in Honor of Elizabeth H. Thomson* (New York: Science History Publications, 1979); Brown, *Benjamin Silliman;* John Patrick Nolan, "Genteel Attitudes in the Formation of the American Scientific Community: The Career of Benjamin Silliman of Yale" (Ph.D. diss., Columbia University, 1978); Guralnick, *Antebellum American College,* 19–21; James H. Cassedy, *Medicine in America: A Short History* (Baltimore: Johns Hopkins University Press, 1991), 25; William G. Rothstein, *American Medical Schools and the Practice of Medicine: A History* (New York: Oxford University Press, 1987), 16; PKR, "Autobiographical Statement," *LL,* 1:8.

14. PKR to Thomas Jefferson, May 21, 1819, *LL,* 1:10; Thomas Jefferson to PKR, June 23, 1819, *LL,* 1:11–12. UVA, chartered in 1819, formally opened in 1825.

15. B. Irvine (Baltimore) to PKR (Williamsburg, Va.), November 9, 1819, WBRP-MITA; Ruby Orders Osborne, *The Crisis Years: The College of William and Mary in Virginia, 1800–1827* (Richmond, Va.: Deitz Press, 1989), 242–55, 260–61; Susan H. Godson et al., *College of William and Mary: A History,* vol. 1: *1693–1888,* 2 vols. (Williamsburg, Va.: King and Queen Press, 1993), 1:204–12.

16. Richard Walsh and William Lloyd Fox, *Maryland: A History, 1632–1974* (Baltimore: Maryland Historical Society, 1974), 123–25; Bernard C. Steiner, *The History of University Education in Maryland, Johns Hopkins University Studies in Historical and Political Science* (Baltimore: Johns Hopkins University Press, 1891), 7–22.

17. Charter reprinted in Godson et al., *William and Mary,* 1:13 Guralnick's *Antebellum American College* mentions William and Mary's academic distinction: "There is no record of how the early program operated, but an actual professorship of mathematics and natural phi-

losophy was established in 1712. Reverend Hugh Jones, educated in England, occupied the position from 1717 to 1729 and was the first professor of science in America" (9); Parker Rouse Jr., *Virginia: The English Heritage in America* (New York: Hastings House, 1976), 103.

18. Drew Gilpin Faust, *A Sacred Circle: The Dilemma of the Intellectual in the Old South, 1840–1860* (Baltimore: Johns Hopkins University Press, 1977), 18, 12. On the quality and character of intellectual life in the antebellum South, see Michael O'Brien, *Conjectures of Order: Intellectual Life and the American South, 1810–1860*, 2 vols. (Chapel Hill: University of North Carolina Press, 2004).

19. Godson et al., *William and Mary*, 1:210, 234.

20. William J. Cooper and Thomas E. Terrill, *The American South: A History* (New York: McGraw Hill, 1996), 245; "Memoir" by WBR, *LL*. 1:15.

21. WBR to JBR, December 22, 1819, *LL*, 1:17.

22. JBR to WBR, February 27, 1823, *LL*, 1:28; JBR to WBR, May 30, 1822, *LL*, 1:22–23.

23. Lester D. Stephens, *Science, Race, and Religion in the American South: John Bachman and the Charleston Circle of Naturalists, 1815–1895* (Chapel Hill: University of North Carolina Press, 2000), chap. 1; Thomas Carey Johnson, *Scientific Interests in the Old South* (New York: Appleton-Century, 1936), 128; see also Greene, *Age of Jefferson*, for his discussion on such "outposts" of science as Charleston. William Gilmore Simms cited in John McCardell, *The Idea of a Southern Nation: Southern Nationalists and Southern Nationalism, 1830–1860* (New York: W. W. Norton, 1979), 155.

24. WBR (Williamsburg, Va.) to JBR (Baltimore), January 19, 1822, WBRP-MITA; WBR to JBR, January 19, 1822, WBRP-MITA.

25. PKR to Thomas Jefferson, January 14, 1824, WBRP-MITA; PKR to Thomas Jefferson, March 14, 1824, cited in G. W. Ewing, *Early Teaching of Science at the College of William and Mary in Virginia* (Williamsburg, Va.: n.p., 1938), 20; see PKR, *An Introduction to the Mathematical Principles of Natural Philosophy* (Richmond, Va.: Shepherd and Pollard, 1822).

26. Godson et al., *William and Mary*, 1:214, 218–20. After the Panic of 1819 William and Mary matriculants fell from an average of eighty-seven students annually to an average of thirty-four students from 1820 to 1825. Some observers in Williamsburg blamed the decline in student enrollments on the decline in the quality of students attending the institution. One resident of the college town stated that "the last session closed with *six* students—and unless some important change is effected in the institution, one Professor after another will probably resign" ("William and Mary College," *Richmond Family Visitor*, July 17, 1824).

27. *LL*, 1:15. The following biographical summaries are based on Robert Rakes Shrock, *Geology at MIT, 1865–1965: A History of the First Hundred Years of Geology at Massachusetts Institute of Technology*, 2 vols. (Cambridge, Mass.: MIT Press, 1977–82), 1:105–7; Patsy Gerstner, *Henry Darwin Rogers, 1808–1866: American Geologist* (Tuscaloosa: University of Alabama Press, 1994); W.S.W. Ruschenberger, *A Sketch of the Life of Robert E. Rogers, M.D., LL.D., with Biographical Notices of His Father and Brothers* (Philadelphia: McCalla and Stavely, 1885); and entries in the *Dictionary of American Biography*.

28. Harris Elwood Starr, "James Blythe Rogers," *Dictionary of American Biography*, 16:99–100.

29. Gerstner, *Henry Darwin Rogers*, 21.

30. Ruschenberger, *Sketch*, 1–35.

31. ES, Preface, *LL*, 1:iii; JBR to WBR, November 9, 1821, *LL*, 1:20.

32. WBR to PKR, November 3, 1826, *LL*, 1:35–36.

33. Gerstner, *Henry Darwin Rogers*, 6; WBR, "Introductory Lecture for the Maryland Institute (January 1827)," *LL*, 1:36–37.

34. WBR to PKR, March 31, 1827, *LL*, 1:41.

35. HDR to PKR, April 20, 1828, *LL*, 1:48.

36. WBR to the governors of the Maryland Institute, April 13, 1828, *LL*, 1:49.

37. Gerstner, *Henry Darwin Rogers*, 7.

38. WBR to PKR, May 19, 1828, *LL*, 1:51; WBR to PKR, June 26, 1828, *LL*, 1:53.

39. HDR to WBR, October 3, 1828, *LL*, 1:60–61.

Chapter Two • Tenure in the Tumult

1. As noted previously, the extant records make it unclear whether WBR ever received a degree from William and Mary; *LL*, 1:62.

2. The entire speech was published as "Address of Professor Rogers" in Williamsburg's *Phoenix Plough-Boy*, November 12, 1828.

3. Historian David Grimsted, in *American Mobbing, 1828–1861* (New York: Oxford University Press, 1998), has argued that the main difference in patterns of social violence between North and South was not in the number of occurrences but in the tendencies or nature of the attacks. "Northern criminals and mobs," he states, "tended to endanger property rather than injure people, while prototypical Southern rioters, like their counterparts in crime, attacked persons more than property. Southern mobs were much likelier to be murderous in intent and/or sadistic in mode than were their Northern counterparts" (86). Along these lines E. Merton Coulter, in *College Life in the Old South* (Athens: University of Georgia Press, 1951), hinted at the North-South differences in student violence by quoting Ralph Waldo Emerson: "The Southerner asks concerning any man, 'How does he fight?' The Northerner asks, 'What can he do?'" (88). Other studies on the character of southern violence include Dickson D. Bruce Jr., *Violence and Culture in the Antebellum South* (Austin: University of Texas Press, 1979); Edward L. Ayers, *Vengeance and Justice: Crime and Punishment in the 19th Century American South* (New York: Oxford University Press, 1984); Elliott J. Gorn, "'Gouge and Bite, Pull Hair and Scratch': The Social Significance of Fighting in the Southern Backcountry," *American Historical Review* 90 (February 1985): 18–43; Bertram Wyatt-Brown, *Honor and Violence in the Old South* (New York: Oxford University Press, 1986); Grady McWhiney, *Cracker Culture: Celtic Ways in the Old South* (Tuscaloosa: University of Alabama Press, 1988); on student uprisings, see Steven J. Novak in *The Rights of Youth: American Colleges and Student Revolt, 1789–1815* (Cambridge: Harvard University Press, 1977); Jennings L. Wagoner, "Honor and Dishonor at Mr. Jefferson's University: The Antebellum Years," *History of Education Quarterly* 26 (Summer 1986): 155–79; Robert F. Pace and Christopher A. Bjornsen, "Adolescent Honor and College Student Behavior in the Old South," *Southern Cultures* (Fall 2000): 9–28; Robert F. Pace, *Halls of Honor: College Men in the Old South* (Baton Rouge, Louisiana State University Press, 2004), chap. 4; Craig Thompson Friend and Lorri Clover, eds., *Southern Manhood: Perspectives on Masculinity in the Old South* (Athens: University of Georgia Press, 2004).

4. Thomas Jefferson, *Notes on the State of Virginia* (1774; rpt., Chapel Hill: University of North Carolina Press, 1955), 162; Jefferson noted that "there must doubtless be an unhappy influence on the manners of our people produced by the existence of slavery among us," especially for children who are "thus nursed, educated, and daily exercised in tyranny" (162). For a case study on the interplay between the culture of slavery and student behavior, see Lewis S. Feuer, "America's First Jewish Professor: James Joseph Sylvester at the University of Virginia," *American Jewish Archives* 36 (November 1984): 152–201. McCardell, *Southern Nationalism,* reviews Jefferson's comparisons between temperaments, North and South, describing northerners as "cool, sober, laborious, independent, interested, chicaning" and southerners as "fiery, voluptuary, indolent, unsteady, generous, candid" (13). Charles Coleman Wall Jr., "Students and the Student Life at the University of Virginia, 1825 to 1861" (Ph.D. diss., University of Virginia, 1979), provides an interpretation of student violence that focuses on the southern code of honor. See also Wagoner, "Honor and Dishonor at Mr. Jefferson's University," 155–79.

5. College of William and Mary, *Minutes of the Faculty,* March 1820.

6. College of William and Mary, *Minutes of the Faculty,* March 10, 1832; see also Manuscript Collections at the College of William and Mary Archive: PKR, Faculty and Alumni Papers; WBR, Faculty and Alumni Papers; and College Papers (1819–35). These collections provide additional background on student culture (i.e., study habits, festivities) and faculty experiences (i.e., teaching, perspectives on discipline).

7. Thomas Cooper quoted in Clement Eaton, *The Freedom of Thought in the Old South* (Durham: Duke University Press, 1940), 27; for an introduction, see the historiographical syntheses on changes in southern political thought in McCardell, *Southern Nationalism;* and William J. Cooper Jr. and Thomas E. Terrill, *The American South: A History* (New York: McGraw Hill, 1996); from a contemporary sociological perspective, Lisa Noel, in *Intolerance: A General Survey* (Montreal: McGill University Press, 1994), argues that, while the contemporary use of the term *intolerance* has come to represent any form of rejection, it has generally meant "the unjustified condemnation of an opinion or behavior" (4). Cooper and Terrill, *American South,* 228–29, 158–59; Eaton, *Freedom of Thought,* 126–31; McCardell, *Southern Nationalism,* 23–24, 30–48.

8. Cooper and Terrill, *American South,* 232–33; see also Larry Tise's *Proslavery: A History of the Defense of Slavery in America, 1701–1840* (Athens: University of Georgia Press, 1987), 70–74; on Dew and the Virginia Convention, see Dickson D. Bruce Jr., *The Rhetoric of Conservatism: The Virginia Convention of 1829–30 and the Conservative Tradition in the South* (San Marino, Calif.: Huntington Library, 1982), 175–93; the proslavery thought of Dew, Nott, Hammond, and others in their southern intellectual context is discussed in Michael O'Brien, *Conjectures of Order: Intellectual Life and the American South, 1810–1860,* 2 vols. (Chapel Hill: University of North Carolina Press, 2004).

9. J. N. Brenaman, *A History of Virginia Conventions* (Richmond, Va.: J. L. Hill Printing Co., 1902), 43–48; David L. Pullman, *The Constitutional Conventions of Virginia from the Foundation of the Commonwealth to the Present Time* (Richmond, Va.: J. T. West, 1901), 63–83; see also Bruce, *Rhetoric of Conservatism,* chap. 2.

10. WBR to HDR, January 2, 1830, box 1, folder 7, WBRP-MITA.

11. WBR to HDR, January 2, 1830, box 1, folder 7, WBRP-MITA.

12. Carl Alfred Lanning Binger, *Revolutionary Doctor: Benjamin Rush, 1746–1813* (New York: W. W. Norton, 1966); JBR to PKR, March 12, 1827, WBRP-MITA; Gerstner, *Henry Darwin Rogers,* 21–22; RER to WBR, December 3, 1832, WBRP-MITA; WBR to HDR, January 2, 1830, box 1, folder 7, WBRP-MITA; David Freeman Hawke, *Benjamin Rush: Revolutionary Gadfly* (Indianapolis: Bobbs-Merrill, 1971); Donald J. D'Elia, *Benjamin Rush: Philosopher of the American Revolution* (Philadelphia: American Philosophical Society, 1974).

13. Francis Lieber quoted in William M. Geer, *Francis Lieber at the South Carolina College* (Raleigh: Print. Shop, North Carolina State College, 1943), 20, 4, 11; Frank Burt Freidel, *Francis Lieber: Nineteenth Century Liberal* (Baton Rouge: Louisiana State University Press, 1948), 129; Thomas Roderick Dew to WBR, November 2, 1835, *LL,* 1:124; WBR to HDR, November 30, 1834, *LL,* 1:112. On Lieber's life and thought, see Charles R. Mack and Henry H. Lesesne, eds., Francis *Lieber and the Culture of Mind* (Columbia: University of South Carolina Press, 2005).

14. WBR to HDR, November 30, 1834, *LL,* 1:113; WBR to HDR, February 11, 1835, *LL,* 1:116–17; *GV,* 762; *Journal of the House of Delegates of the Commonwealth of Virginia . . . 1835* (Richmond, Va.: Samuel Shepherd, 1835), 76.

15. Faust, *Sacred Circle,* 7–14; see also Wyatt-Brown, *Honor and Violence,* 44–51; Virginia's internal sectionalism discussed in William Shade, *Democratizing the Old Dominion: Virginia and the Second Party System, 1824–1861* (Charlottesville: University Press of Virginia, 1996); Alison Goodyear Freehling, *Drift toward Dissolution: The Virginia Slavery Debate of 1831–32* (Baton Rouge: Louisiana State University Press, 1982); "Memorial to the Legislature of the Commonwealth of Virginia, Adopted at Full Meeting of the Citizens of Kanawha, [Document No. 8]," *Journal of the House of Delegates of Virginia, Session 1841–42* (Richmond, Va.: Samuel Shepard, 1841), 6, quoted in Sean Patrick Adams, "Old Dominions and Industrial Commonwealths: The Political Economy of Coal in Virginia and Pennsylvania, 1810–1875" (Ph.D. diss., University of Wisconsin, Madison, 1999), 160–61; see also "The Inequality of Representation in the General Assembly of Virginia: A Memorial to the Legislature of the Commonwealth of Virginia, Adopted at Full Meeting of the Citizens of Kanawha," *West Virginia History* 25 (July 1964): 283–98; Sean Patrick Adams, "Partners in Geology, Brothers in Frustration: The Antebellum Geological Surveys of Virginia and Pennsylvania," *Virginia Magazine of History and Biography* 106 (Winter 1998): 5–34.

16. William Ernst, "William Barton Rogers: Antebellum Virginia Geologist," *Virginia Cavalcade* 24 (Summer 1974): 15; George Summers to WBR, February 15, 1836, Board of Public Works Collection, Virginia State Library.

17. WBR to HDR, March 8, 1838, *LL,* 1:152–53.

18. WBR to HDR, April 1, 1839, *LL,* 1:163; Gerstner, *Henry Darwin Rogers,* 50–54. HDR requested an annual budget of five thousand dollars for the survey, and the Pennsylvania legislature provided sixty-four hundred dollars for the project. Robert Rakes Shrock, *Geology at MIT, 1865–1965: A History of the First Hundred Years of Geology at Massachusetts Institute of Technology,* 2 vols. (Cambridge, Mass.: MIT Press, 1977–82), 1:166–67.

19. WBR to HDR and RER, January 18, 1841, *LL,* 1:181.

20. Judge J. F. May to WBR, March 16, 1841, *LL,* 1:183; WBR to RER, September 11, 1841, *LL,* 1:191–92.

21. Joseph Henry on the issue of slavery quoted in Moyer, *American Scientist,* 199; letter

of recommendation from Joseph Henry, July 6, 1835, *LL*, 1:126; for references to Rogers's appointment, see University of Virginia, *Minutes of the Board of Visitors*, July 8, 1835; Joseph C. Cabell to James Madison, July 25, 1835, WBR Faculty/Alumni File, College of William and Mary.

22. Thomas Jefferson to William Roscoe, December 27, 1820, quoted in Wayne Hamilton Wiley, "Academic Freedom at the University of Virginia: The First Hundred Years—From Jefferson through Alderman" (Ph.D. diss., University of Virginia, 1973), 75. Quotations from Virginius Dabney, *Mr. Jefferson's University: A History* (Charlottesville: University Press of Virginia, 1981), 8–9; Philip A. Bruce, *History of the University of Virginia, 1819–1919: The Lengthened Shadow of One Man*, 5 vols. (New York: Macmillan, 1920), 2:34–35, 298; more recent works include Cameron Addis, *Jefferson's Vision for Education, 1760–1845* (New York: Peter Lang, 2003) Jennings L. Wagoner Jr., *Jefferson and Education* (Charlottesville: Thomas Jefferson Foundation, 2004); and the republication of Dumas Malone's classic *The Sage of Monticello* (Charlottesville: University of Virginia Press, 2005). Additional sources relevant to this period can be found in the University Papers (1835–53); and WBR Papers located at the UVA Archives.

23. "University of Virginia," *Watchman of the South*, August 5, 1841, quoted in Feuer, "Sylvester," 156–57; Bruce, *University of Virginia*, 3:79. Much has been written about the Sylvester controversy. This and the following account on the mathematician is largely derived from Feuer's, "Sylvester"; as well as Raymond Clare Archibald, "Unpublished Letters of James Joseph Sylvester and Other New Information concerning His Life and Work," *Osiris* 1 (January 1936): 85–154; R. C. Yates, "Sylvester at the University of Virginia," *American Mathematical Monthly* 44 (1937): 194–201; "Sylvester in Virginia," *Mathematical Intelligencer* 9 (1987): 3–19; I. M. James, "James Joseph Sylvester, F.R.S. (1814–1897)," *Notes and Records of the Royal Society of London* 51 (July 1997): 247–61; Karen Hunger Parshall, *James Joseph Sylvester: Life and Works in Letters* (Oxford: Oxford University Press, 1998), 12–13. For a comprehensive, biographical treatment of Sylvester's life, see Karen Hunger Parshall, *James Joseph Sylvester: Jewish Mathematician in a Victorian World* (Baltimore: Johns Hopkins University Press, 2006).

24. WBR to RER, September 11, 1841, *LL*, 1:192; for an extended discussion of Sylvester's initial reception, see Feuer, "Sylvester," 154 ff.; and Parshall, *James Joseph Sylvester: Jewish Mathematician.*

25. After leaving Virginia, Sylvester recounted being "in a state of utter . . . despondency" to Harvard mathematician Benjamin Peirce in the following letters located in the Benjamin Peirce Papers, Houghton Library, Harvard University, Cambridge: J. J. Sylvester to Benjamin Peirce, February 28, 1843; J. J. Sylvester to Benjamin Peirce, May 19, 1843; J. J. Sylvester to Benjamin Peirce, May 22, 1843; J. J. Sylvester to Benjamin Peirce, June 11, 1843; for further commentary on the mathematician's departure and later career, see Feuer, "Sylvester," 158–69; Archibald, "Unpublished Letters of James Joseph Sylvester"; and James, "James Joseph Sylvester, F.R.S."

26. John A. G. Davis, *An Exposition of the Proceedings of the Faculty of the University of Virginia in Relation to the Recent Disturbances at That Institution* (Charlottesville: J. Alexander, 1836), reprinted in December 16, 1836, *LL*, 1:139. The chairmanship was equivalent to the presidency of the University of Virginia at the time. This incident is also discussed in Feuer,

"Sylvester," 175–77; Feuer adds an account of the horsewhipping of classics professor Gessner Harrison by his students in 1839. WBR to Brothers in Philadelphia, November 16, 1840, *LL*, 1:176–177; see also Gorn, "Fighting in the Southern Backcountry," 18–43.

27. Student unrest described in Bruce, *University of Virginia*, 3:113–14; WBR to HDR, April 4, 1845, *LL*, 1:247.

28. JBR to Brothers [WBR and RER], January 10, 1845, box 2, folder 22, WBRP-MITA; WBR to HDR, April 29, 1845, *LL*, 1:249.

29. *Journal of the House of Delegates of Virginia. Session, 1844–45* (Richmond, Va.: Samuel Shepherd, 1844), 38, 40, 43.

30. *Journal of the House of Delegates of Virginia, Session 1844–45*, 105; WBR, "Report from the Committee of Schools and Colleges against the Expediency of Withdrawing the Fifteen Thousand Dollars Annuity from the University [Document No. 41]," *Journal of the House of Delegates, Session 1844–45*, reprinted in *LL*, 1:400.

31. WBR, "Report from the Committee of Schools and Colleges," *LL*, 1:401, 408–9, 411.

32. William R. Johnson to WBR, February 14, 1845, *LL*, 1:241; WBR to William R. Johnson, March 15, 1845, *LL*, 1:242; WBR to HDR, April 5, 1846, *LL*, 1:264.

33. WBR to J. C. Cabell, March 14, 1848, *LL*, 1:280–281; J. C. Cabell to WBR, April 2, 1848, *LL*, 1:282; WBR to HDR, March 21, 1848, *LL*, 1:282–283; WBR to HDR, April 29, 1845, *LL*, 1:250.

34. Quotation on the Savages in George Stillman Hillard, *Memoir of the Hon. James Savage, LL.D., Late President of the Massachusetts Historical Society* (Boston: John Wilson and Son, 1878), 17, 32; WBR to HDR, March 13, 1846, *LL*, 1:259; James Savage to his daughter and her husband, November 23, 1852, reprinted in Emma Rogers, ed., *Letters of James Savage to His Family* (Boston: n.p., 1906), 167.

35. WBR to HDR, April 29, 1845, *LL*, 1:250.

Chapter Three · From Soils to Species

1. WBR often grumbled about the difficulty of establishing his scientific career while at UVA: "We who are in collegiate harness" he wrote, "may well envy the lot of those happy fellows who, free from all such restraints, can go whithersoever, the love of research impels, and can devote all their hours of vigorous thought to extending the boundaries of knowledge" (*LL*, 1:227). Nicholas Jardine, James A. Secord, and Emma C. Spary, eds., *Cultures of Natural History* (New York: Cambridge University Press, 1996), include a cultural survey of approaches to nineteenth-century natural history; Michael Dettelbach, "Humboldtian Science," in Jardine et al., *Natural History*, 288–89.

2. On the state of stratigraphy in America during this period, see Patsy A. Gerstner, "Henry Darwin Rogers and William Barton Rogers on the Nomenclature of the American Paleozoic Rocks," in Cecil J. Schneer, ed., *Two Hundred Years of Geology in America* (New Hanover: University Press of New England, 1979), 175–86; Tocqueville, *Democracy in America* (London, 1835–40), cited in Hugo A. Meier, "Technology and Democracy, 1800–1860," *Mississippi Valley Historical Review* 43 (1957): 624.

3. The two most important influences on WBR's geological thought were Patrick, his father, and Henry, his brother. See "Address of Professor Rogers," *Phoenix Plough-Boy*, Novem-

ber 12, 1828; Gerstner, *Henry Darwin Rogers,* 13–27. For a concise history of American geology, see Leonard G. Wilson, "The Emergence of Geology as a Science in the United States," *Journal of World History* 10 (1967): 416–37; William Browning, "The Relation of Physicians to Early American Geology," *Annals of Medical History* 3 (1931): 547–60, 565; A. D. Bache and HDR, "Analysis of Some Coals of Pennsylvania," *Journal of the Academy of Natural Sciences of Philadelphia* 7 (1834): 158–77. See also Hugh Richard Slotten, *Patronage, Practice, and the Cu of American Science: Alexander Dallas Bache and the U.S. Coast Survey* (New York: Cambridge University Press, 1994); Henry S. Patterson, *Memoir of the Life and Scientific Labors of Samuel George Morton* (Philadelphia: Lippincott, Grambo and Co., 1854); George B. Wood, *A Biographical Memoir of Samuel George Morton* (Philadelphia: T. K. and P. G. Collins, 1853); Charles D. Meigs, *A Memoir of Samuel George Morton, M.D., Late President of the Academy of Natural Sciences of Philadelphia* (Philadelphia: T. K. and P. G. Collins, 1851); Edmund Berkeley and Dorothy Smith Berkeley, *George William Featherstonhaugh: The First U.S. Government Geologist* (Tuscaloosa: University of Alabama Press, 1988).

4. On Ruffin and southern agriculture, see Richard C. Sheridan, "Mineral Fertilizers in Southern Agriculture," in James X. Corgan, ed., *The Geological Sciences in the Antebellum South* (Tuscaloosa: University of Alabama Press, 1982), 73–82; Avery O. Craven, *Edmund Ruffin, Southerner: A Study in Secession* (New York: D. Appleton and Co., 1932); David F. Allmendinger Jr., *Ruffin: Family and Reform in the Old South* (New York: Oxford University Press, 1990), 36–49.

5. Faust, *Sacred Circle,* 11; see also Alfred Glaze Smith, *Economic Readjustment of an Old Cotton State: South Carolina, 1820–1860* (Columbia: University of South Carolina Press, 1958); Avery Odell Craven, *Soil Exhaustion as a Factor in the Agricultural History of Virginia and Maryland, 1606–1860* (Urbana: University of Illinois Press, 1926), provides a discussion on the migration out of Virginia; for a sampling of Rogers's research characteristic of his useful arts approach (authored and coauthored) during the 1830s and 1840s, see "Observations and Queries Respecting Artesian Wells," *Farmers' Register* 7 (December 1834): 451–55; HDR, "Observations on the Geology of the Western Peninsula of Upper Canada, and the Western Part of Ohio," *Proceedings of the American Philosophical Society* 2 (1842): 120–25; HDR, "Theory of Earthquake Action," *American Journal of Science* 45 (1843): 341–47; HDR, "On the Phenomena of the Great Earthquakes . . . to Elucidate Several Points in Geological Dynamics," *Proceedings of the American Philosophical Society* 3 (1843): 64–67; HDR, "On the Geological Age of the White Mountains," *American Journal of Science* 1 (1846): 411–21; RER, "On the Absorption of Carbonic Acid Gas by Liquids," *American Journal of Science* 6 (1848): 96–110.

6. WBR, "On the Discovery of Green Sand," in *GV,* 3 and 8; WBR, "Chemical Analysis of Shells," *Farmers' Register* 1 (March 1834): 591.

7. WBR, "Apparatus for Analyzing Marl and the Carbonates in General," in *GV,* 11.

8. WBR, "Apparatus for Analyzing Calcareous Marl and other Carbonates," *American Journal of Science* 27 (1835): 299–300; WBR, "A Self-Filling Syphon for Chemical Analysis," *American Journal of Science* 27 (1835): 302.

9. WBR, "Further Observations on the Green Sand," in *GV,* 12; WBR, "On the Discovery of Green Sand," in *GV,* 6.

10. WBR, "Chemical Analysis of Shells," 589; WBR, "Further Observations on the Green Sand," in *GV,* 15.

11. Berkeley and Berkeley, *George William Featherstonhaugh,* 112–13; Peter A. Browne to John Floyd, governor of Virginia, September 30, 1833, reprinted in *GV,* 750. Additional sources that discuss the Virginia survey include Michele L. Aldrich and Alan E. Leviton, "William Barton Rogers and the Virginia Geological Survey, 1835–1842," in Corgan, *Geological Sciences,* 83–104; William Ernst, "William Barton Rogers: Antebellum Virginia Geologist," *Virginia Cavalcade* 24 (1974): 13–20; on the conduct and expectations of the survey, see Benjamin R. Cohen, "Surveying Nature: Environmental Dimensions of Virginia's First Scientific Survey, 1835–1842," *Environmental History* 11 (2006): 37–69.

12. Browne to Floyd, September 30, 1833, reprinted in *GV,* 752, 750; see also Jennings L. Wagoner Jr., "Honor and Dishonor at Mr. Jefferson's University: The Antebellum Years," *History of Education Quarterly* 26 (Summer 1986): 155–79. Wagoner discusses the place of religion in the Old South: "Not until late in the antebellum period did evangelical Christianity severely alter the dominant characteristics that defined the ideal southern gentleman. The anticlerical tradition associated with Jefferson and other southern gentry under the spell of the rationalism of the Enlightenment, coupled with planter resistance to church power and patronage, served to limit the status of ministers and diminish the appeal of the church in much of southern society. . . . [O]nly a fifth to a third of all southern whites before the Civil War were churchgoers" (161–62).

13. *Report from the Select Committee of the General Assembly of Virginia . . . Praying for a Geological Survey of the State, with a View to the Discovery and Development of Its Geological and Mineral Resources,* reprinted in *GV,* 761, 759, 755; for further discussion about state surveys of the period, see Anne Marie Millbrooke, "State Geological Surveys of the Nineteenth Century" (Ph.D. diss., University of Pennsylvania, 1981).

14. WBR to HDR, February 11, 1835, *LL,* 1: 116–17.

15. *GV,* 24.

16. *GV,* 26, 27, 543.

17. *GV,* 41–51.

18. *GV,* 156, 134, 281–82.

19. *GV,* 91.

20. Some classic and popular field guides used in the nineteenth century include John Lettsome, *Naturalist's and Traveller's Companion* (London: C. Dilly, 1774); Edward Donovan, *Instructions for Collecting and Preserving Various Subjects of Natural History* (London: n.p., 1794); George Graves, *Naturalist's Pocket-Book and Tourist's Companion* (London: Longman, 1818). For a discussion on these and other guides used by geologists of the period, see Anne Larsen, "Equipment for the Field," in Jardine et al., *Natural History,* 358–77.

21. *GV,* 193.

22. WBR and RER, "An Account of Some New Instruments and Processes for the Analysis of the Carbonates," *American Journal of Science* 46 (April 1844): 347; *GV,* 143; on a comparative perspective on funding for state geological surveys, see Gerstner, *Henry Darwin Rogers,* 50, 54.

23. HDR to WBR, April 10, 1836, *LL,* 1:130; WBR to HDR, December 22, 1840, Geological Survey Papers, Library of Virginia. I am indebted to Aldrich and Leviton, "Virginia Geological Survey," for calling my attention to these letters. This collection has an extensive repository of correspondence between Rogers and his associates on the survey. Some of these

materials have been effectively mined to analyze the organization of the survey in Cohen, "Surveying Nature."

24. See Gerstner, "Nomenclature of the American Paleozoic Rocks," 175–86.

25. Gerstner, "Nomenclature of the American Paleozoic Rocks," interprets the failure of the Rogers system of stratigraphy as stemming from the brothers' ineffective efforts at publicizing the numbering arrangement. By Hall's desire for priority I refer to his interest in claiming credit for establishing the names used in American stratigraphy.

26. For a detailed description of the paper and its impact on the scientific community of the period, see Patsy Gerstner, "A Dynamic Theory of Mountain Building: Henry Darwin Rogers, 1842," *Isis* 66 (1975): 26–37; Gerstner, however, gives William less credit than he merited in the creation of the theory. For one thing William's name was listed first in the publication. He often published with his brothers and regularly traded first authorship depending on their involvement in the research. Moreover, Gerstner overlooked the significant amount of interest William had in natural philosophy. The "dynamic theory" likely emerged from William's natural philosophy research, an area of research to which Henry gave far less attention.

27. Gabriel Gohau, *A History of Geology* (New Brunswick, N.J.: Rutgers University Press, 1991), 106–7; Mott T. Greene, *Geology in the Nineteenth Century: Changing Views of a Changing World* (Ithaca: Cornell University Press, 1982), 19. The term *uniformitarianism* was coined after *Huttonianism* by William Whewell in 1832. For a classic discussion and commentary on Lyell and uniformitarianism, see Martin J. S. Rudwick, "Uniformity and Progression: Reflections on the Structure of Geological Theory in the Age of Lyell," in Duane H. D. Roller, ed., *Perspective in the History of Science and Technology* (Norman: University of Oklahoma Press, 1971), 209–37; on Lyell's experiences with and observations of mid-nineteenth America, its society, culture, and scientific community, see Leonard G. Wilson, *Lyell in America: Transatlantic Geology, 1841–1853* (Baltimore: Johns Hopkins University Press, 1998).

28. WBR and HDR, "On the Physical Structure of the Appalachian Chain as Exemplifying the Laws Which Have Regulated the Elevation of Great Mountain Chains Generally," reprinted in *GV,* 624, 642; Greene, *Geology in the Nineteenth Century,* provides a survey of the North American debates over mountain formation theory by such figures as James Hall, James Dwight Dana, and Joseph LeConte.

29. Gerstner, "Dynamic Theory," 26, 35–37; HDR to RER, November 5, 1848, *LL,* 1:293: some of the Continental geologists whom Henry Rogers referred to included "Argo, Pouillet, Dumas, Pentland, De Verneuil, Elie de Beaumont, Count D' Archiac, Valenciennes, and others."

30. For examples of Rogers's works in natural philosophy and chemistry reflecting his useful arts approach, see "On the Transporting Power of Currents," *American Journal of Science* 5 (1848): 115–16; "Some Experiments on Sonorous Flames, with Remarks on the Primary Source of Their Vibrations," *American Journal of Science* 26 (1858): 1–15; "On the Formation of Rotating Rings by Air and Liquids under Certain Conditions of Discharge," *American Journal of Science* 26 (1858): 246–58. During the 1850s and 1860s Rogers continued to present and publish on geological topics. A few representative studies include "On the Origin and Accumulation of the Protocarbonate of Iron in Coal Measures," *Proceedings of the Boston Society of Natural History* 5 (1856): 283–88; "Discovery of Paleozoic Fossils in Eastern Massachu-

setts," *American Journal of Science* 22 (1856): 296–98; "On the Group of Rocks Constituting the Base of the Paleozoic Series in the United States," *Proceedings of the Boston Society of Natural History* 7 (1861): 394–95.

31. WBR and HDR, "Experimental Enquiry into Some of the Laws of the Elementary Voltaic Battery," *American Journal of Science* 27 (1835): 44.

32. Rogers and Rogers, "Voltaic Battery," 44–45, 52.

33. Rogers and Rogers, "Voltaic Battery," 60.

34. For a brief survey of natural philosophy texts of the early to mid-nineteenth century, see Edward W. Stevens Jr., *The Grammar of the Machine: Technical Literacy and Early Industrial Expansion in the United States* (New Haven: Yale University Press, 1995), 65–71. Most science and mathematics texts of this period, argues Stevens, were written by only a few individuals.

35. WBR, *An Elementary Treatise on the Strength of Materials* (Charlottesville, Va.: Tompkins and Noel, 1838), 3.

36. WBR, *Strength of Materials*, 7, 39.

37. WBR, *Elements of Mechanical Philosophy* (Boston: Thurston, Torrey, and Emerson, 1852), 3.

38. WBR, *Mechanical Philosophy*, 6.

39. WBR, *Address before the Lyceum of Natural History of Williams College, August 14, 1855* (Boston: T. R. Marvin and Son, 1855), 9–10.

40. WBR, *Address before the Lyceum*, 12.

41. Edward Lurie, *Louis Agassiz: A Life in Science* (Baltimore: Johns Hopkins University Press, 1988), 31–71.

42. On the reception of Darwinism in America, see George Daniels, *Darwinism Comes to America* (Waltham: Blaisdell Publishing Co., 1968), 22–29; Edward J. Pfeifer, "United States," in Thomas F. Glick, ed., *The Comparative Reception of Darwinism* (Austin: University of Texas Press, 1974), 176–81; David L. Hull, *Darwin and His Critics: The Reception of Darwin's Theory of Evolution by the Scientific Community* (Cambridge: Harvard University Press, 1973). General and comparative histories also include Ronald Numbers, *Darwinism Comes to America* (Cambridge: Harvard University Press, 1998); and Ronald L. Numbers and John Stenhouse, eds., *Disseminating Darwinism: The Role of Place, Race, Religion, and Gender* (New York: Cambridge University Press, 2000).

43. A. Hunter Dupree, *Asa Gray: American Botanist, Friend of Charles Darwin* (Baltimore: Johns Hopkins University Press, 1988), 110.

44. On the reception of evolution in France, see Robert E. Stebbins, "France," in Glick, *Comparative Reception*, 117–63; additional works on Asa Gray and Louis Agassiz include Edward Lurie, *Nature and the American Mind: Louis Agassiz and the Culture of Science* (New York: Science History Publications, 1974); Ian F. A. Bell, "Divine Patterns: Louis Agassiz and American Men of Letters, Some Preliminary Explorations," *Journal of American Studies* 10 (1976): 349–81; Ralph W. Dexter, "The Impact of Evolutionary Theories on the Salem Group of Agassiz Zoologists (Morse, Hyatt, Packard, Putnam)," *Essex Institute Historical Collections* 115 (1979): 144–71; Paul Jerome Croce, "Probabilistic Darwinism: Louis Agassiz vs. Asa Gray on Science, Religion, and Certainty," *Journal of Religious History* 22 (1998): 35–58; Kenneth W. Hermann, "Shrinking from the Brink: Asa Gray and the Challenge of Darwinism, 1853–1868"

(Ph.D. diss., Kent State University, 1999); on Agassiz's brand of creationism, see David K. Nartonis, "Louis Agassiz and the Platonist Story of Creation at Harvard, 1795–1846," *Journal of the History of Ideas* 66 (2005): 437–49.

45. Lurie, *Agassiz,* 126–27; Dupree, *Gray,* 123. On the relationship between personality and scientific achievement in Louis Agassiz's career, see James R. Jackson and William C. Kimler, "Taxonomy and the Personal Equation: The Historical Fates of Charles Girard and Louis Agassiz," *Journal of the History of Biology* (Netherlands) 32 (1999): 509–55.

46. Lurie, *Agassiz,* 252–302.

47. Elizabeth Cary Agassiz, ed., *Louis Agassiz: His Life and Correspondence* (New York: Houghton Mifflin, 1885), 437, 411.

48. Edward Lurie, "Louis Agassiz and the Idea of Evolution," *Victorian Studies* (September 1959): 92; HDR to WBR, December 23, 1859, *LL,* 2:17; WBR, "Literature [Review of *On the Origin of Species*]," *Boston Courier,* March 5, 1860.

49. BSNH, *Proceedings* 7 (1861): 168.

50. BSNH, *Proceedings* 7 (1861): 173; American Academy of Arts and Sciences, *Proceedings* 4 (1860): 360.

51. BSNH, *Proceedings* 7 (1861): 231, 232.

52. BSNH, *Proceedings* 7 (1861): 232, 233–35. One observer wrote to Rogers on what he believed to be the general sentiment after the final debate: "I have been much interested— somewhat instructed—and highly amused at last by the late discussions opened by Agassiz to say as much as possible about Darwin and closed by him with the 'desire to say as little as possible.' Time settles all things and Darwin can take of himself. Meantime I enjoyed your surprise to find Agassiz so ingeniously turn the tables on you about the shallow seas . . . the last geological idea expressed by Agassiz which startled and astonished everybody" (C. F. Winslow to WBR, April 5, 1860, WBRP-MITA); see also Nathaniel Southgate Shaler, *The Autobiography of Nathaniel Southgate Shaler* (Boston: Houghton Mifflin, 1909), 105.

53. BSNH, *Proceedings* 7 (1861): 234–35; WBR to HDR, February 20, 1860, WBRP-MITA.

54. BSNH, *Proceedings* 7 (1861): 244–45, 274.

55. BSNH, *Proceedings* 7 (1861): 168, 173, 231, 232, 234–35, 244–45, 274; American Academy of Arts and Sciences, *Proceedings* 4 (1860): 360; on the relationship between Agassiz and the Socratic method, see Lane Cooper, *Louis Agassiz as Teacher: Illustrative Extracts on His Method of Instruction* (Ithaca: Comstock Publishing Co., 1917), 3–4. A. E. Verrill recounted his years as student and later as assistant to Agassiz and said that under the great zoologist and geologist "any independence of action or of thought (if expressed) is nearly impossible" (A. E. Verrill to WBR, Dec. 7, 1868, WBRP-MITA).

56. Shaler, *Autobiography,* 105; C. F. Winslow to WBR, April 5, 1860, WBRP-MITA; Jules Marcou, *Life, Letters, and Works of Louis Agassiz* (New York: Macmillan, 1896), 108–9.

57. Gray and a cohort of evolutionists, including but not limited to Alpheus Hyatt, Edward D. Cope, and Othaniel C. Marsh, were active in the diffusion of evolutionary thought in the mid- to late nineteenth century. Rogers's scientific research, meanwhile, rarely dealt directly with Darwinism or other branches of evolution after the debates with Agassiz. See Pfeifer, "United States," 181–206; Dexter, "Impact of Evolutionary Theories," 148. On Alexan-

der Agassiz's ideas about evolution, see David Dobbs, *Reef Madness: Alexander Agassiz, Charles Darwin, and the Meaning of Coral* (New York: Pantheon, 2005).

Chapter Four · Advancing and Diffusing

1. Hunter Dupree, *Science in the Federal Government: A History of Policies and Activities* (Baltimore: Johns Hopkins University Press, 1986), 91–114; see also Robert V. Bruce, *The Launching of American Science, 1846–1876* (New York: Knopf, 1987), chap. 12. The scholarship on the professionalization of science has focused, as historian Nathan Reingold has observed, on "a story of how full-time professionals necessarily and inevitably supplanted amateurs, however talented and devoted" ("Definitions and Speculations: The Professionalization of Science in America in the Nineteenth Century," in Alexandra Oleson and Sanborn C. Brown, eds., *The Pursuit of Knowledge in the Early American Republic: American Scientific and Learned Societies from Colonial Times to the Civil War* [Baltimore: Johns Hopkins University Press, 1976], 33). Professionalization occurred in response to increasingly complex bodies of scientific knowledge, to the desire among scientists to communicate this new knowledge to others participating in the increasing specialization, and to the desire for institutionalized self-government that would regulate standards of scientific activity. According to George H. Daniels, historian of science, "the emergence of a community of such professionals was the most significant development in nineteenth century American science" ("The Process of Professionalization in American Science: The Emergent Period, 1820–1860," in Nathan Reingold, ed., *Science in America since 1820* [New York: Science History Publications, 1976], 63).

2. On Bache and the survey, see Hugh Richard Slotten, *Patronage, Practice, and the Culture of American Science: Alexander Dallas Bache and the U.S. Coast Survey* (New York: Cambridge University Press, 1994), 77–111.

3. Bruce Sinclair, *Philadelphia's Philosopher Mechanics: A History of the Franklin Institute, 1824–1865* (Baltimore: Johns Hopkins University Press, 1974), 250–51; Bruce, *American Science,* 252–53. The Columbian Institute lasted from 1816 to 1838; the American Geological Society was organized by Benjamin Silliman in 1819.

4. Sally Kohlstedt, *The Formation of the American Scientific Community: The American Association for the Advancement of Science, 1846–1860* (Urbana: University of Illinois Press, 1976), 66; WBR to Benjamin Silliman, January 18, 1845, AAGN MSS, Academy of Natural Sciences of Philadelphia, cited in Sally Kohlstedt, "The Geologists' Model for National Science, 1840–1847," *Proceedings of the American Philosophical Society* 118 (1974): 192.

5. *American Journal of Science* 41 (October 1841): 259; *American Journal of Science* 43 (October 1842): 176; *Reports of the First, Second, and Third Meetings of the Association of American Geologists and Naturalists* (Boston: Gould, Kendall and Lincoln, 1843), 69.

6. *American Journal of Science* 47 (October 1844): 111–12; *American Journal of Science* 41 (October 1841): 242.

7. *Reports of the First, Second, and Third Meetings,* 72.

8. WBR to J. W. Bailey, October 22, 1843, cited in Kohlstedt, "Model for National Science," 188.

9. Kohlstedt, *American Scientific Community,* 59–77.

10. HDR to RER, November 5, 1848, *LL*, 1:293. The invitations from London and Copenhagen were sent in 1844.

11. *Proceedings of the American Association for the Advancement of Science: First Meeting Held at Philadelphia, September, 1848* (Philadelphia: John C. Clark, 1849), 62; *American Journal of Science* 47 (October 1844): 106.

12. *Reports of the First, Second, and Third Meetings,* 68.

13. Kohlstedt, "Model for National Science," 192–93; WBR to Benjamin Silliman, January 18, 1845, AAGN MSS, Academy of Natural Sciences of Philadelphia, cited in Kohlstedt, "Model for National Science," 192.

14. *Proceedings of the American Association for the Advancement of Science,* 2, 12, 26.

15. WBR to his brothers, September 21, 1849, *LL*, 1:305.

16. WBR to HDR, July 13, 1849, *LL*, 1:299; WBR to HDR, August 19, 1849, *LL*, 1:302; WBR to HDR, October 5, 1849, *LL*, 1:309.

17. Kohlstedt, *American Scientific Community,* 78–99.

18. HDR to WBR, May 16, 1848, *LL*, 1:288. By "popularization" of science I refer to the Lazzaroni's equating of practical forms of science with charlatanism. See Lilian B. Miller, *The Lazzaroni: Science and Scientists in Mid-Nineteenth Century America* (Washington, D.C.: Smithsonian Institution Press, 1972); as well as Presidential Addresses by members of the Lazzaroni printed in the *Proceedings* of the American Association for the Advancement of Sciences, particularly during the 1850s.

19. Kohlstedt, *American Scientific Community,* 100–107.

20. Kohlstedt, *American Scientific Community,* 97.

21. Kohlstedt, *American Scientific Community,* 97; see also John D. Holmfeld, "From Amateurs to Professionals in American Science: The Controversy over the Proceedings of an 1853 Scientific Meeting," *Proceedings of the American Philosophical Society* 114 (February 1970): 22–36.

22. Kohlstedt, *American Scientific Community,* 161.

23. Hugh R. Slotten, "Science, Education, and Antebellum Reform: The Case of Alexander Dallas Bache," *History of Education Quarterly* 31 (1991): 323–42.

24. See Bruce, *Modern American Science,* chap. 9.

25. Sources that discuss the origins and activities of the Lazzaroni include Mark Beach, "Was There a Scientific Lazzaroni?" in George H. Daniels, ed., *Nineteenth Century American Science: A Reappraisal* (Evanston, Ill.: Northwestern University Press, 1972); Bruce, *Modern American Science,* 263–66; and Thomas L. Haskell, *The Emergence of Professional Social Science: The American Social Science Association and the Nineteenth Century Crisis of Authority* (Chicago: University of Chicago Press, 1977), 68–74.

26. Bruce, *Modern American Science,* 265. Sally Kohlstedt identifies John Warner and Daniel Vaughn as the authors of the appellation "Washington-Cambridge Clique" (Kohlstedt, *American Scientific Community,* 156). Miller, in *The Lazzaroni,* lists Asa Gray, WBR, CWE, and Matthew Fountaine Maury as the main opponents; during this controversy at the AAAS, however, Rogers reported being joined or supported by a different group of Lazzaroni dissenters that included Ormsby M. Mitchel, Chester Dewey, Edward Hitchcock, and J. W. Bailey (WBR to HDR, September 1, 1856, WBRP-MITA); Kohlstedt adds that J. Lawrence Smith and William Hackley aided Rogers as well.

27. Kohlstedt, *American Scientific Community,* 182.

28. WBR to A. D. Bache, April 2, 1854, *LL,* 1:339; WBR to J. Lovering, June 30, 1854, WBRP-MITA; Kohlstedt, *American Scientific Community,* 181.

29. Kohlstedt, *American Scientific Community,* 183.

30. WBR to HDR, September 1, 1856, WBRP-MITA.

31. WBR to HDR, September 1, 1856, WBRP-MITA. On the scientists Rogers considered allies, see Philip Stanley Shoemaker, "Stellar Impact: Ormsby Macknight Mitchel and Astronomy in Antebellum America" (Ph.D. diss., University of Wisconsin, Madison, 1991); Martin Brewer Anderson, *Sketch of the Life of Professor Chester Dewey, D.D., LL.D., Late Professor of Chemistry and Natural History in the University of Rochester* (Albany: n.p., 1868); Florence Beckwith, *Early Botanists of Rochester and Vicinity and the Botanical Section* (Rochester: Rochester Academy of Science, 1912); Edward Singleton Holden, *Biographical Memoir of William H. C. Bartlett, 1804–1893* (Washington, D.C.: National Academy of Science, 1911).

32. *Proceedings of the American Association for the Advancement of Science* (Philadelphia: John C. Clark, 1857), 231; WBR to Lorin Blodgett, September 1, 1856, WBRP-MITA; Kohlstedt, *American Scientific Community,* 184; WBR to HDR, September 1, 1856, WBRP-MITA.

33. A. D. Bache, *Anniversary Address before the American Institute of the City of New York, at the Tabernacle, October 28th, 1856* (New York: Pudney and Russell, 1857), cited in Richard J. Storr, *The Beginnings of Graduate Education in America* (New York: Arno Press, 1969), 91.

Chapter Five · Thwarted Reform

1. CWE, "The New Education," *Atlantic Monthly* 23 (1869): 203–20; on college-level science before the nineteenth century, see Theodore Hornberger, *Scientific Thought in the American Colleges, 1639–1800* (Austin: University of Texas Press, 1945); on science and curricular changes during the antebellum period, see Guralnick, *Antebellum American College;* Frederick Rudolph, *Curriculum: A History of the American Undergraduate Course of Study since 1636* (San Francisco: Jossey-Bass, 1977), 55–98; Christopher J. Lucas, *American Higher Education: A History* (New York: St. Martin's Press, 1994), 104–37; Scott L. Montgomery, *Minds for the Making: The Role of Science in American Education, 1750–1990* (New York: Guilford Press, 1994); Roger Geiger, ed., *The American College in the Nineteenth Century* (Nashville: Vanderbilt University Press, 2000); John Thelin, *A History of American Higher Education* (Baltimore: Johns Hopkins University Press, 2004).

2. Martin Kaufman, *American Medical Education: The Formative Years, 1765–1910* (Westport, Conn.: Greenwood Press, 1976); Ronald Numbers, ed., *The Education of American Physicians* (Berkeley: University of California Press, 1980); Lamar Riley Murphy, *Enter the Physician: The Transformation of Domestic Medicine, 1760–1860* (Tuscaloosa: University of Alabama Press, 1991); Thomas Neville Bonner, *Becoming a Physician: Medical Education in Britain, France, Germany, and the United States, 1750–1945* (New York: Oxford University Press, 1995); Chandos Michael Brown, *Benjamin Silliman: A Life in the Young Republic* (Princeton: Princeton University Press, 1989); Leonard G. Wilson, ed., *Benjamin Silliman and His Circle: Studies on the Influence of Benjamin Silliman on Science in America: Prepared in Honor of Elizabeth H. Thomson* (New York: Science History Publications, 1979); see also Guralnick, *Antebellum American College,* 19–21.

3. Sally Kohlstedt, "Parlors, Primers, and Public Schooling: Education for Science in Nineteenth Century America," *Isis* 81 (1990): 425–45. Kohlstedt argues that "we are looking for national commitment to science in the wrong place . . . if we simply try to find outstanding scientists and piece together their experiences" (424). Guralnick, *Antebellum American College,* 26, 27, 41–42, 35–36. Lyceums and institutes of science and technology also developed before, during, and after the 1820s: West Point (1802), Norwich (1820), Gardiner Lyceum (1823), Rensselaer School (1824), Franklin Institute (1824), Virginia Military Institute (1839), the Citadel (1843), and the U.S. Naval Academy (1845). A number of short-lived institutions also appeared, such as Polytechnic College of Pennsylvania (1853), Brooklyn Polytechnic Institute (1855), Cooper Union (1859), and schools in Cleveland, Ohio (1857), and Glenmore, New York (1859); see also Terry S. Reynolds, "The Education of Engineers in America before the Morrill Act of 1862," *History of Education Quarterly* 32 (1992): 459–82.

4. Ruby Orders Osborne, *The Crisis Years: The College of William and Mary in Virginia, 1800–1827* (Richmond, Va.: Deitz Press, 1989), 313.

5. *Laws and Regulations of the College of William and Mary in Virginia* (Richmond, Va.: Thomas W. White, 1830), 13; WBR, *An Elementary Treatise on the Strength of Materials* (Charlottesville: Tompkins and Noel, 1838); WBR, *Elements of Mechanical Philosophy* (Boston: Thurston, Torrey, and Emerson, 1852); Harold Smith, *The Society for the Diffusion of Useful Knowledge, 1826–1846: A Social and Bibliographical Evaluation* (London: Vine Press, 1974).

6. *Laws and Regulations,* 12; John White Webster, *A Manual of Chemistry on the Basis of Professor Brande's* (Boston: Richard and Lord, 1828), v–vi; William Thomas Brande, *A Manual of Chemistry* (London: J. Murray, 1819).

7. RER to HDR, December 6, 1829, *LL, 1:*77–78.

8. WBR to HDR, December 6, 1828, *LL, 1:*68–69.

9. UVA, *Catalogue* (1835–36), 15; UVA, *Catalogue* (1843–44), 14. The courses differed markedly between those offered by Rogers's predecessor in 1834–35 and those offered immediately after Rogers's arrival; the variety of courses taught and texts used by Rogers are cited in the *Catalogue* of the university from 1835 to 1853.

10. UVA, *Minutes from the Board of Visitors Meetings,* August 13, 1836; the Civil Engineering program description first appeared in the UVA, *Catalogue* (1836–37); UVA, *Catalogue* (1837–38), 15; Philip A. Bruce, *History of the University of Virginia, 1819–1919: The Lengthened Shadow of One Man,* 5 vols. (New York: Macmillan, 1920), 2:126.

11. UVA, *Catalogue* (1849–50), 17.

12. For a discussion on methods of instruction in nineteenth-century American higher education, see Linda Armstrong Chisholm, "The Art of Undergraduate Teaching in the Age of the Emerging University" (Ph.D. diss., Columbia University, 1982); Caroline Winterer, in *The Culture of Classicism: Ancient Greece and Rome in American Intellectual Life, 1780–1910* (Baltimore: Johns Hopkins University Press, 2002), discusses the resistance that even reform-minded classicists met from their colleagues as they attempted to teach more about "worlds" than "words."

13. "Original Papers in Relation to a Course of Liberal Education," *American Journal of Science and Arts* 25 (1829): 197–351. Various interpretations of the Yale Report of 1828 are discussed in Jack C. Lane, "The Yale Report of 1828 and Liberal Education: A Neorepublican Manifesto," *History of Education Quarterly* 27 (1987): 325–38. Classicists, of course, were

hardly a monolithic group and differed among themselves on how and what to teach their students. Most focused primarily on teaching grammar, while others called attention to ancient culture. Those attending to culture debated the merits of using translations versus the original texts. Still others favored the scholarship on the classical languages emanating from Germany, while others frowned on it. Even questions about the superiority of Latin versus Greek entered into their debates. Whatever their differences, however, they all benefited from the impact that Yale had in reasserting the significance of classical studies. On classicists of the period, see Winterer, *Culture of Classicism.*

14. "Original Papers in Relation to a Course of Liberal Education," *American Journal of Science and Arts* 25 (1829): 300–301; Walter B. Kolesnik, *Mental Discipline in Modern Education* (Madison: University of Wisconsin Press, 1958), provides a still useful account of the origins and development of faculty psychology.

15. Bruce, *University of Virginia,* 3:249.

16. Bruce, *University of Virginia,* 3:249; Bruce Sinclair, *Philadelphia's Philosopher Mechanics: A History of the Franklin Institute, 1824–1865* (Baltimore: Johns Hopkins University Press, 1974), 230; WBR, "For the Establishment of a School of Arts, Franklin Institute, Philadelphia," folder 14b, box 1, WBRP-MITA.

17. *For the Establishment of a School of Arts. Memorial of the Franklin Institute, of the State of Pennsylvania, for the Promotion of the Mechanic Arts, to the Legislature of Pennsylvania* (Philadelphia: J. Crissy, 1837), 7–9.

18. *School of Arts,* 3, 9–10.

19. *School of Arts,* 9, 11–12. "Who can doubt," asked Rogers, "that the mechanical, manufacturing, and agricultural classes of the community would derive the highest advantages from the establishment of an institution directed by these views" (11). For a case study on the transition from the apprenticeship system to factory labor, see Paul G. Faler, *Mechanics and Manufacturers in the Early Industrial Revolution: Lynn, Massachusetts, 1780–1860* (Albany: State University of New York Press, 1981).

20. Sinclair, *Philosopher Mechanics,* 134.

21. On the development of the scientific disciplines in the early to mid-nineteenth century undergraduate curriculum, see Guralnick, *Antebellum American College.* Chisholm, in "Undergraduate Teaching," argues that "for good or ill, recitation was synonymous with nineteenth century classroom teaching" (31); see her discussion on the recitation, the laboratory, and mid-nineteenth-century attempts at instructional reform.

22. Guralnick, *Antebellum American College,* 127–30; CWE, "New Education," 203–20; see also Roger Geiger, "The Rise and Fall of Useful Knowledge: Higher Education for Science, Agriculture, and the Mechanic Arts, 1850–1875," in Geiger, ed., *The American College in the Nineteenth Century* (Nashville: Vanderbilt University Press, 2000), 153–68.

23. Mary Ann James, "Engineering an Environment for Change: Bigelow, Peirce, and Early Nineteenth-Century Practical Education at Harvard," in Clark A. Elliot and Margaret W. Rossiter, eds., *Science at Harvard University: Historical Perspectives* (Bethlehem, Pa.: Lehigh University Press, 1992), 67, 69; CWE, "New Education," 209; Tenney L. Davis, "Eliot and Storer: Pioneers in the Laboratory Teaching of Chemistry," *Journal of Chemical Education* 6 (1929): 870.

24. WBR to HDR, October 3 and 13, 1847, *LL,* 1:274. According to Chisholm, "Under-

graduate Teaching," laboratories for student use were not a central part of college science un-til the postbellum period. At Yale students did not use laboratories until the twentieth cen-tury, in part, because nonundergraduate programs such as the Sheffield Scientific School de-layed developments at the college. Harvard laboratories for instruction remained a scattered and unofficial part of the undergraduate program until the last quarter of the nineteenth cen-tury. At Amherst facilities for laboratories appeared in the 1890s. Columbia made its first offi-cial declaration of support for the laboratory in 1897.

25. HDR to WBR, March 8, 1846, *LL,* 1:256–58; WBR to HDR, March 13, 1846, *LL,* 1:259–60, 420–27.

26. WBR to HDR, March 13, 1846, *LL,* 1:259–62.

27. WBR, "Plan for a Polytechnic School in Boston (1846)," *LL,* 1:420, 421.

28. WBR, "Plan for a Polytechnic School in Boston (1846)," *LL,* 1:422, 425–26, 426–27. Virginia instituted full coeducation in 1970.

29. Charles C. Smith, "Memoir of John Amory Lowell, LL.D.," *Massachusetts Historical Society, Proceedings* (January 1898): 118–19. I am indebted to Tachikawa, "Two Sciences and Religion," 128–30, for his discussion on the similarities between the two plans.

30. HDR to WBR, July 15, 1859, *LL,* 2:11; WBR to HDR, March 20, 1860, *LL,* 2:30; Mar-garet Rossiter, "Louis Agassiz and the Lawrence Scientific School" (B.S. thesis, Harvard Uni-versity, 1966).

31. On Wayland, see Frederick Rudolph, *The American College and University: A History* (New York: Knopf, 1962), 237–40; Frederick Rudolph, *Curriculum: A History of the Ameri-can Undergraduate Course of Study since 1636* (San Francisco: Jossey-Bass, 1977), 109–12; Christopher J. Lucas, *American Higher Education: A History* (New York: St. Martin's Press, 1994), 136–37.

32. HDR to WBR, December 22, 1849, *LL,* 1:311.

33. Francis Wayland, *Thoughts on the Present Collegiate System in the United States* (Boston: Gould, Kendall and Lincoln, 1842); Francis Wayland, *A Memoir of the Life and Labors of Fran-cis Wayland, D.D., LL.D., Late President of Brown University* (New York: Sheldon, 1867), 92; WBR to HDR, April 18, 1850, *LL,* 1:313; Rudolph, *American College and University,* 238.

Chapter Six · Instituting a New Education

1. During this period, it should be noted, many American colleges had expanded in vary-ing degrees their scientific offerings. What is most impressive is the gradual but steady change that occurred in the faculty and curriculum of traditional institutions. In 1828, for instance, Williams College had only one lecturer for science, but by 1830 it had four of seven faculty members involved in science instruction. The University of Pennsylvania underwent a simi-lar change in faculty distribution, with three of six members teaching math or science by 1836. Most colleges had only one professor of scientific studies in 1800. But by the mid-nineteenth century almost all colleges had positions for distinguished professors of math and science, with many colleges having more than half of their faculty in these disciplines. See Guralnick, *Antebellum American College,* ix.

For Rogers, of course, the quantitative change mattered little if not coupled with a qual-itative change. Having more students learning outmoded science by way of recitation offered no reason to celebrate for reformers like Rogers. He wanted an independent program of science that would be free to enact the kinds of quantitative and qualitative changes in higher education that he believed would be necessary for instruction to keep abreast of research. To his mind such an institute would need to be comprehensive (i.e., covering a variety of practical and theoretical topics) as well as specialized (i.e., depth of scientific studies beyond that of traditional programs).

2. Rudolph, *American College and University,* 238; RER to WBR, January 7, 1833, *LL,* 1:01; James L. Morrison Jr., "Educating the Civil War Generals: West Point, 1833–1861," *Military Affairs* 38 (1974): 108–11; other works that discuss the early years at West Point include Stephen Ambrose, *Duty Honor, Country: A History of West Point* (Baltimore: Johns Hopkins University Press, 1999); Sidney Forman, *West Point: A History of the United States Military Academy* (New York: Columbia University Press, 1950); R. Ernest Dupuy, *Sylvanus Thayer: Father of Technology in the United States* (New York: United States Military Academy, 1958).

3. In large measure WBR's views of military academies were expressed earlier in his opposition to the military system of the Virginia Military Institute and in his defense of the UVA's mission (intellectual freedom). Forman, *West Point,* 167.

4. Palmer C. Ricketts, *History of the Rensselaer Polytechnic Institute, 1824–1894* (New York: Wiley, 1895), 6–7; Rudolph, *Curriculum,* 63; other works on RPI include Ray Palmer Baker, "Rensselaer Polytechnic Institute and the Beginnings of Science in the United States," *Scientific Monthly* 19 (October 1924): 337–56; Palmer C. Ricketts, *Rensselaer Polytechnic Institute: A Short History* (Troy, N.Y.: RPI Press, 1930); Palmer C. Ricketts, *Rensselaer Polytechnic Institute* (Troy, N.Y.: RPI Press, 1933); Samuel Rezneck, *Education for a Technological Society: A Sesquicentennial History of Rensselaer Polytechnic Institute* (Troy, N.Y.: RPI Press, 1968).

5. Rezneck, *Rensselaer Polytechnic Institute,* 111–31; Benjamin Franklin Greene's ideas on the "True Polytechnic" was published as *The Rensselaer Polytechnic Institute: Its Reorganization in 1849–50; Its Condition at the Present Time: Its Plans and Hopes for the Future* (Troy, N.Y.: D. H. Jones, 1855).

6. Storr, *Graduate Education,* 116; Andrew D. White, "Scientific and Industrial Education in the United States," *Popular Science Monthly* 5 (June 1874), reprinted in Carroll W. Pursell Jr., ed., *Readings in Technology and American Life* (New York: Oxford University Press, 1969), 182. According to Yale president Noah Porter in *The American College and the American Public* (New Haven: C. C. Chatfield, 1870), the University of Michigan had been cited as "a decisive argument in favor of radical reform," although his inspection of the curriculum showed that it was "on the whole very old-fashioned and conservative in its most distinguishing features . . . [which] does not differ materially from that of any college which is provided with a scientific and technological school" (22). Robert Silverman and Mark Beach, "A National University for Upstate New York," *American Quarterly* 22 (1970): 701–13; S. Edward Warren, *Notes on Polytechnic or Scientific Schools in the United States: Their Nature, Position, Aims and Wants* (New York: Wiley, 1866). A discussion on science and engineering programs can be found in Terry S. Reynolds, "The Education of Engineers in America before the Morrill Act of 1862," *History of Education Quarterly* 32 (1992): 459–82.

7. Several institutes of the kind that interested Rogers had appeared in Germany, Sweden, and Switzerland during the early to mid-nineteenth century. For a brief survey of the

practical and theoretical values of these European institutions, see Rolf Torstendahl, "The Transformation of Professional Education in the Nineteenth Century," in Sheldon Rothblatt and Bjorn Wittrock, eds., *The European and American University since 1800: Historical and Sociological Essays* (New York: Cambridge University Press, 1993), 109–41. Torstendahl argues that European technical education emerged for two basic reasons: "demand from the State for a labour force" and the "industrial economy and . . . capitalist agriculture" (125). The French polytechnic schools, more so than others in Europe, deeply influence the documents that Rogers would later prepare for the founding of MIT. See *Objects and Plan of an Institute of Technology* (Boston: J. Wilson, 1861); Frederick B. Artz, *The Development of Technical Education in France, 1500–1850* (Cleveland: Society for the History of Technology, 1966), 145. For an alternate interpretation of Rogers's European influences, see Stratton and Mannix, *Mind and Hand,* 435–36, 540–41. They offer evidence that suggests Germany's Karlsruhe influenced his ideas about museum organizing, rather than scientific instruction. Many classic works in history of education by such scholars as Frederick Rudolph, Hugh Hakins, and others have emphasized German traditions of scientific studies in American higher education and have not fully considered the French influences. See, for example, Lawrence Veysey's *The Emergence of the American University* (Chicago: University of Chicago Press, 1965), 125–33; for a recent analysis of the making of modern European higher education, see Walter Ruegg, *Universities in the Nineteenth and Early Twentieth Centuries, 1800–1945* (New York: Cambridge University Press, 2004).

8. Report of the Society of Arts, December 1, 1864, *LL,* 2:216–17; Artz, *Technical Education in France,* 247–53; Charles R. Day, *Education for the Industrial World: The Ecole d'Arts et Metiers and the Rise of French Industrial Engineering* (Cambridge, Mass.: MIT Press, 1987), 12–15. Jean-Baptist Dumas, one of the original faculty members at the Ecoles Central, described his approach to teaching theory as it applied to practical studies. "My intention," he declared, "has not been to describe the practice of the arts, but to clarify the theory of them" (Dumas cited in John Hubbel Weiss, *The Making of Technological Man: The Social Origins of French Engineering Education* [Cambridge, Mass.: MIT Press, 1982], 116).

9. WBR to HDR, September 16, 1851, *LL,* 1:319.

10. "Governor's Address," *Acts and Resolves Passed by the General Court of Massachusetts in the Year 1859* (Boston: Secretary of the Commonwealth, 1859), 488–89; *LL,* 2:2.

11. George S. Boutwell, *Thoughts on Educational Topics and Institutions* (Boston: Phillips, Sampson, 1859), 333, 337; "Senate No. 86," *Documents Printed by Order of the Senate of the Commonwealth of Massachusetts during the Session of the General Court, A.D. 1859* (Boston: n.p., 1859); Tachikawa, "Two Sciences and Religion," 251.

12. "House No. 260," *Documents Printed by Order of the House of Representatives of the Commonwealth of Massachusetts during the Session of the General Court, A.D. 1859* (Boston: Dutton and Wentworth, 1859), 4, 5, 9, 10.

13. *LL,* 2:3; "House No. 260," *Documents Printed by Order of the House of Representatives of the Commonwealth of Massachusetts during the Session of the General Court, A.D. 1859* (Boston: Dutton and Wentworth, 1859), 5–6, 8; James P. Munroe, "The Beginning of the Massachusetts Institute of Technology," *Technology Quarterly* 1 (May 1888): 289.

14. *LL,* 2:3.

15. "House No. 13," *Documents Printed by Order of the House of Representatives of the Com-*

monwealth of Massachusetts during the Session of the General Court, A.D. 1860 (Boston: Dutton and Wentworth, 1860), reprinted in *LL*, 2:406.

16. "House No. 13," *Documents Printed by Order of the House of Representatives of the Commonwealth of Massachusetts during the Session of the General Court, A.D. 1860* (Boston: Dutton and Wentworth, 1860), reprinted in *LL*, 2:416. While Stratton and Mannix, *Mind and Hand,* argue that the museum was the foremost part of the three-part MIT plan, Rogers's *Address before the Lyceum of Natural History of Williams College, August 14, 1855* (Boston: T. R. Marvin and Son, 1855), suggests that he favored the school over the other two parts.

17. "House No. 119," *Documents Printed by Order of the House of Representatives of the Commonwealth of Massachusetts during the Session of the General Court, A.D. 1860* (Boston: Dutton and Wentworth, 1860), 3; Tachikawa, "Two Sciences and Religion," 248.

18. "House No. 260," *Documents Printed by Order of the House of Representatives of the Commonwealth of Massachusetts during the Session of the General Court, A.D. 1859* (Boston: Dutton and Wentworth, 1859), 11; Paul Goodman, "Ethics and Enterprise: The Values of a Boston Elite, 1800–1860," *American Quarterly* 18 (Fall 1966): 437–51; Ronald Story, *The Forging of an Aristocracy: Harvard and the Boston Upper Class, 1800–1870* (Middletown: Wesleyan University Press, 1980); Betty Farrell, *Elite Families: Class and Power in Nineteenth-Century Boston* (Albany: State University of New York Press, 1993). A classic study of the "Boston Brahmin" class can be found in David Tyack, *George Ticknor and the Boston Brahmins* (Cambridge: Harvard University Press, 1967), 173–83. Tyack suggests that this New England "caste" had difficulty translating its "economic, social and intellectual authority into political power. . . . Consequently Ticknor and a number of his conservative friends sought to bypass parties and legislature and to influence the course of the nation in other ways. They sought to control institutions—schools, churches, libraries, the legal system, the republic of letters—which would stabilize society" (183). While a dated interpretation, the thrust of Tyack's depiction of the so-called Brahmins applies to the milieu in which Rogers proposed ideas about education for the industrial classes.

19. WBR to HDR, March 30, 1860, *LL*, 2:29.

20. *Objects and Plan of an Institute of Technology* (Boston: J. Wilson, 1861); *An Account of the Proceedings Preliminary to the Organizations of the Massachusetts Institute of Technology* (Boston: J. Wilson and Son, 1861). Each of the three parts, whether he intended them to or not, paralleled the scientific, professional, and educational values he had sustained across his career. The Society of Arts satisfied his research and professional interests, while the museum and science programs followed from his educational reform ambitions.

21. *Objects and Plan,* 6, 8.

22. *Objects and Plan,* 9, 10–11.

23. *Objects and Plan,* 13, 15.

24. *Objects and Plan,* 17.

25. *Objects and Plan,* 21–22, 27.

26. *Objects and Plan,* 22–23.

27. *Objects and Plan,* 25, 28.

28. *Account of the Proceedings,* 4.

29. Samuel Prescott, *When MIT Was "Boston Tech," 1861–1916* (Cambridge, Mass.: MIT Press, 1954), 29–30; *Account of the Proceedings,* 4–5.

30. *Account of the Proceedings,* 5–7.

31. William P. Blake to WBR, December 3, 1860, WBRP-MITA; James E. Olivier to WBR, January 12, 1861, WBRP-MITA; James Ritchie to WBR, December 6, 1860, WBRP-MITA; newspaper account on Peirce extracted in *LL,* 2:62. More on Blake in David B. Dill Jr. "William Phipps Blake: Yankee Gentleman and Pioneer Geologist of the Far West," *Journal of Arizona History* 32 (1991): 385–412.

32. *Account of the Proceedings,* 17, 23.

33. WBR to HDR, February 5, 1861, *LL,* 2:67; WBR to HDR, March 19, 1861, *LL,* 2:73.

34. M. D. Ross, *Estimate of the Financial Effect of the Proposed Reservation of Back-Bay Lands* (Boston: J. Wilson and Son, 1861), 5.

35. Governor Andrew to WBR, March 9, 1861, *LL,* 2:75; WBR, *Report of the Joint Standing Committee of the Massachusetts Legislature of 1861 on the Memorial of the Associated Institutions of Science and Art* (March 19, 1861), reprinted in *LL,* 2:424–29; WBR to HDR, February 18, 1861, *LL,* 2:70. *LL,* 2:78.

36. WBR to Governor Andrew, March 28, 1861, Governor John Andrew Papers, Massachusetts Historical Society, cited in Tachikawa, "Two Religions," 259.

Chapter Seven · Convergence of Interests

1. William K. Scarborough, ed., *The Diary of Edmund Ruffin: Toward Independence, October, 1856–April, 1861* (Baton Rouge: Louisiana State University Press, 1972), 588; David F. Allmendinger Jr., *Ruffin: Family and Reform in the Old South* (New York: Oxford University Press, 1990), 171.

2. WBR to HDR, April 16, 1862, *LL,* 2:116; Robert V. Bruce, *Lincoln and the Tools of War* (Indianapolis: Bobbs-Merrill, 1956); Robert V. Bruce, *The Launching of Modern American Science* (New York: Knopf, 1987), chap. 20; A. Hunter Dupree, *Science in the Federal Government: A History of Policies and Activities* (Baltimore: Johns Hopkins University Press, 1986), 120–48; Bruce, *Launching of American Science,* chap. 23. The heaviest wartime grief felt by Rogers and his family was most likely over the loss of James Savage Jr. Basic information on the members and organization of NAS, see the following holdings in the National Academy of Sciences Archives: A. D. Bache Member File; Joseph Henry Member File; Benjamin Peirce Member File; RER Member File; and WBR Member File. "An Act to Incorporate the Massachusetts Institute of Technology, and to Grant Aid to Said Institute and to the Boston Society of Natural History," *Acts and Resolves Passed by the General Court of Massachusetts in the Year 1861* (Boston: Secretary of the Commonwealth, 1861), 492–95; *LL,* 2:78.

3. "An Act for the Inspection of Gas Meters, the Protection of Gas Consumers and the Protection and Regulation of Gas Light Companies," *Acts and Resolves Passed by the General Court of Massachusetts in the Year 1861* (Boston: Secretary of the Commonwealth, 1861), 480–85; WBR was appointed in June, 1861; *LL,* 2:90; WBR to Governor John A. Andrew, June 20, 1861, *LL,* 2:90–91; WBR to HDR, June 25, 1861, *LL,* 2:91–92. Rogers also conducted geological work for the state. When Governor Andrew asked him review a proposal for surveying select coal regions in the state, Rogers replied in characteristic useful arts fashion: "If there it be thought expedient for the state to engage in such an investigation, it should not I think

be content with a merely local and partial exploration, but should do the work so thoroughly as to decide the question as to the extent and availableness of the coal of this region once and for all" (WBR to Governor Andrews, December 27, 1862, WBRP-MITA)

4. WBR to HDR, June 25, 1861, *LL,* 2:92.

5. WBR to HDR, June 25, 1861, *LL,* 2:92; WBR to HDR, January 20, 1863, *LL,* 2:147; WBR to HDR, July 7, 1861, *LL,* 2:93.

6. WBR to HDR, September 17, 1861, *LL,* 2:96.

7. WBR to HDR, September 17, 1861, *LL,* 2:96; WBR to HDR, December 1, 1861, *LL,* 2:102–3; WBR to HDR, April 28, 1862, *LL,* 2:117–18.

8. WBR to HDR, April 28, 1862, *LL,* 2:119; WBR to HDR, October 13, 1862, *LL,* 2:133.; WBR to HDR, March 31, 1863, *LL,* 2:157; *LL,* 2:173; WBR to HDR, January 19, 1864, *LL,* 2:185; WBR to Governor John A. Andrew, February 1, 1864, *LL,* 2:187.

9. "Resolves Concerning the Universal Exposition at Paris," *Acts and Resolves Passed by the General Court of Massachusetts in the Year 1866* (Boston: Secretary of the Commonwealth, 1866), 317–18; *LL,* 2:264; Robert Brain, *Going to the Fair: Readings in the Culture of Nineteenth Century Exhibitions* (Cambridge: Whipple Museum of the History of Science, 1993), 9–18, 32, 45–46.

10. WBR to James Savage Sr., June 25, 1867, *LL,* 2:272; Brain, *Going to the Fair,* 32, 46; WBR to James Savage Sr., July 27, 1867, *LL,* 2:275. Rogers became ill and never completed the report on the Paris Exposition. For correspondence on the stalled report, see CWE to WBR, July 12, 1869, WBRP-MITA; F. H. Storer to WBR, July 17, 1869, WBRP-MITA; WBR to Governor William Claflin, January 29 and 31, 1870, WBRP-MITA.

11. WBR to HDR, November 17, 1863, *LL,* 2:180; additional research studies conducted by WBR during this period include: "[On] the Frozen Well of Brandon, Vermont," *Boston Society of Natural History Proceedings* 9 (1862): 72–81; "Electrical Illumination at Boston Photometrically Measured," *American Journal of Science* 36 (1863): 307–8; "An Account of Apparatus and Processes for Chemical and Photometrical Testing of Illuminating Gas," *British Association Report* 34 (1864): 39–40.

12. Thomas L. Haskell, *The Emergence of Professional Social Science: The American Social Science Association and the Nineteenth-Century Crisis of Authority* (Urbana: University of Illinois Press, 1977); Mary O. Furner, *Advocacy and Objectivity: A Crisis in the Professionalization of American Social Science, 1865–1905* (Lexington: University Press of Kentucky, 1975); Dorothy Ross, *The Origins of American Social Science* (New York: Cambridge University Press, 1991); Lawrence Goldman, in "Exceptionalism and Internationalism: The Origins of American Social Science Reconsidered," *Journal of Historical Sociology* (UK) 11 (March 1998), argues that the nineteenth-century American social science community displayed "a self-conscious internationalism" (1). Rogers's own internationalism supports Lawrence's argument, which differs from Ross's American exceptionalism thesis.

13. Haskell, *Emergence of Professional Social Science,* 115.

14. National Academy of Sciences, *A History of the First Half-Century of the National Academy of Sciences, 1863–1913* (Washington, D.C.: National Academy of Sciences, 1913); Rexmond C. Cochrane, *The National Academy of Sciences: The First Hundred Years, 1863–1963* (Washington, D.C.: National Academy of Sciences, 1978).

15. Dupree, *Science in the Federal Government,* 135–41; Cochrane, *National Academy of Sciences,* 58–63; WBR, "Memoranda of the Meeting for Organising the National Academy of Sciences," WBR, Member File, NAS Archives.

16. WBR, "Memoranda of the Meeting;" WBR to HDR, March 17, 1863, *LL,* 2:154–55; WBR to HDR, April 28, 1863, *LL,* 2:161–62; Donald Fleming, *John William Draper and the Religion of Science* (Philadelphia: University of Pennsylvania Press, 1950); George Frederick Barker, *Memoir of John William Draper, 1811–1882. By George F. Barker. Read before the National Academy, April 21, 1886* (Washington, D.C.: n.p., 1886).

17. WBR, "Memoranda of the Meeting;" WBR to HDR, April 28, 1863, *LL,* 2:162; two very rich collections that contain references to the professionalization of science during this period are located in the Houghton Library at Harvard University: Benjamin Peirce Papers and A. D. Bache Papers.

18. Benjamin Peirce to A. D. Bache, May 27, 1863, cited in Dupree, *Science in the Federal Government,* 144.

19. J. S. Newberry to WBR, September 5, 1865, *LL,* 2:248; Dupree, *Science in the Federal Government,* 147.

20. M. D. Ross to WBR, June 20, 1861, WBRP-MITA; "An Act in Addition to an Act to Incorporate the Massachusetts Institute of Technology," *Acts and Resolves Passed by the General Court of Massachusetts in the Year 1862* (Boston: Secretary of the Commonwealth, 1862), 93; the most comprehensive and recent study on MIT's early years is Stratton and Mannix, *Mind and Hand.* For other studies on the Institute's history, see Samuel Prescott, *When MIT Was "Boston Tech," 1861–1913* (Cambridge, Mass.: MIT Press, 1954); Richard Rakes Shrock, *Geology at MIT, 1865–1965: A History of the First Hundred Years of Geology at Massachusetts Institute of Technology,* 2 vols. (Cambridge, Mass.: MIT Press, 1977–82); Silas W. Holman, "Massachusetts Institute of Technology," in George Gary Bush, ed., *History of Higher Education in Massachusetts* (Washington, D.C.: GPO, 1891), 280–319; James P. Munroe, "The Beginning of the Massachusetts Institute of Technology," *Technology Quarterly* 1 (May 1888): 285–97; the following unpublished papers located in the MIT Archives have called my attention to relevant primary sources: Sarah Slaughter, "The Origins of MIT," (MS, MIT, 1980); Carolyne Kirdahy, "The Morrill Land Grant Act and the Massachusetts Institute of Technology" (MS, MIT, 1989).

21. Prescott, *Boston Tech,* 35–37; Loretta H. Mannix, "Communications to the Society of Arts at Its Regular Meetings Beginning with the Meeting of December 12, 1862" (MS, MIT Archives, 1979); see also the MIT Annual Reports of the 1870s for accounts of the Society of Arts meetings. Although the society claimed to award noteworthy innovations prizes or honors, Rogers did not want his or the Institute's name used for endorsements. See WBR to Mr. Peylis, February 6, 1865, WBRP-MITA; WBR to Dr. Whelpley, June 6, 1867, WBRP-MITA.

22. Roger L. Williams, *The Origins of Federal Support for Higher Education: George W. Atherton and the Land-Grant College Movement* (University Park: Pennsylvania State University Press, 1991); on the relationship between the state, universities, and the Morrill Act of 1862, see Mark R. Nemec, *Ivory Towers and Nationalist Minds: Universities, Leadership, and the Development of the American State* (Ann Arbor: University of Michigan Press, 2006).

23. Louis Agassiz to Governor John A. Andrew, December 16, 17, and 22, 1862, John Andrew Papers, Massachusetts Historical Society.

24. Louis Agassiz to Governor John A. Andrew, December 22, 1862, John Andrew Papers, Massachusetts Historical Society.

25. Governor John A. Andrew to WBR, December 22 and 30, 1862, WBRP-MITA; WBR to William Walker, May 4, 1863, *LL*, 2:163–64. In the May 4 letter Rogers provided an extract of his reply to Governor Andrew.

26. Record of BSNH meeting, February 18, 1859, WBRP-MITA.

27. Linda Armstrong Chisholm, "The Art of Undergraduate Teaching in the Age of the Emerging University" (Ph.D. diss., Columbia University, 1982), 242–43, 250; Mary P. Winsor, *Reading the Shape of Nature: Comparative Zoology at the Agassiz Museum* (Chicago: University of Chicago Press, 1991), 6, 12; on Rogers's views of museum work and its "statistical" goals, see WBR, *Address before the Lyceum of Natural History of Williams College, August 14, 1855* (Boston: T. R. Marvin and Son, 1855), 9–10.

28. See WBR, *Address before the Lyceum,* 9–10; Winsor, *Reading the Shape of Nature;* Mary P. Winsor, "Agassiz's Notions of a Museum: The Vision and the Myth," in Michael T. Ghiselin and Alan E. Leviton, eds., *Cultures and Institutions of Natural History* (San Francisco: California Academy of Sciences, 2000), 249–71.

29. Thomas Hill to Governor John A. Andrew, December 24, 1862, John Andrew Papers, Massachusetts Historical Society; Samuel Eliot Morison, *Three Centuries of Harvard, 1636–1936* (Cambridge: Harvard University Press, 1936), 306.

30. "Governor's Address," *Acts and Resolves Passed by the General Court of Massachusetts in the Year 1863* (Boston: Secretary of the Commonwealth, 1863), 618.

31. WBR to HDR, March 31, 1863, *LL*, 2:157; WBR to HDR, March 17, 1863, *LL*, 2:153; *An Account of the Proceedings Preliminary to the Organizations of the Massachusetts Institute of Technology* (Boston: J. Wilson and Son, 1861). A number of state documents provide greater detail about the deliberations among state leaders regarding the merger as well as relevant petitions submitted by the Massachusetts Boards of Agriculture and Trade. Those documents include "Resolve Authorizing Certain Expenditures by the Committee on an Agricultural College," *Acts and Resolves Passed by the General Court of Massachusetts in the Year 1863* (Boston: Secretary of the Commonwealth, 1863), 560; "Senate No. 108," *Documents Printed by Order of the Senate of the Commonwealth of Massachusetts during the Session of the General Court, A.D. 1863* (Boston: n.p., 1863), 4–5; "Senate No. 108," *Documents Printed by Order of the Senate of the Commonwealth of Massachusetts during the Session of the General Court, A.D. 1863* (Boston: n.p., 1863), 18; "Senate No. 108," *Documents Printed by Order of the Senate of the Commonwealth of Massachusetts during the Session of the General Court, A.D. 1863* (Boston: n.p., 1863), 10.

32. "Senate No. 108," *Documents Printed by Order of the Senate of the Commonwealth of Massachusetts during the Session of the General Court, A.D. 1863* (Boston: n.p., 1863), 16, 30; WBR to William Walker, May 4, 1863, *LL*, 2:163; "An Act to Provide for the Reception of a Grant of Congress, and to Create a Fund for the Promotion of Education in Agriculture and the Mechanic Arts," *Acts and Resolves Passed by the General Court of Massachusetts in the Year 1863* (Boston: Secretary of the Commonwealth, 1863), 480–81; Harold Whiting Cary, *The University of Massachusetts: A History of One Hundred Years* (Amherst: University of Massachusetts Press, 1962).

33. "An Act in Addition to the Act to Incorporate the Massachusetts Institute of Tech-

nology," *Acts and Resolves Passed by the General Court of Massachusetts in the Year 1863* (Boston: Secretary of the Commonwealth, 1863), 496–97.

34. "Governor's Address," *Acts and Resolves Passed by the General Court of Massachusetts in the Year 1863* (Boston: Secretary of the Commonwealth, 1863), 727–28; *Public Documents of Massachusetts: Being the Annual Reports of Various Public Offices and Institutions for the Year 1864*, vol. 1: *Public Document Number 4, Annual Report of the Massachusetts Board of Agriculture* (Boston: William White, 1865), 44.

35. *LL*, 2:155; Prescott, *Boston Tech*, 40; "William Johnson Walker," *Dictionary of American Biography*, 22 vols. (New York: Charles Scribner's Sons, 1943), 19:366.

36. *Scope and Plan of the School of Industrial Science, Massachusetts Institute of Technology* (Boston: J. Wilson and Son, 1864), 3; Silas, "Massachusetts Institute of Technology," 294–95.

37. Silas, "Massachusetts Institute of Technology," 295–96; *Scope and Plan;* see also *First Annual Catalogue of the Officers and Students and Programme of the Course of Instruction of the School of the Massachusetts Institute of Technology, 1865-6* (Boston: J. Wilson and Son, 1865). The Institute did not receive permission from the state to grant degrees until 1868.

38. Silas, "Massachusetts Institute of Technology," 295–96. See the MIT *Catalogue* for 1865–66: "A high value is set upon the educational effect of laboratory practice, in the belief that such practice trains the senses to observe with accuracy, and the judgment to rely with confidence on the proof of actual experiment" (27).

39. Silas, "Massachusetts Institute of Technology," 296–97.

40. Silas, "Massachusetts Institute of Technology," 296.

41. WBR to HDR, June 15, 1863, *LL*, 2:166; WBR to HDR, January 19, 1864, *LL*, 2:186. While working on the plan for the Institute, Rogers also received help from a colleague in Paris who helped make comparisons between the idea of MIT and schools in Europe. (W. G. Preston to WBR, March 26, 1864, WBRP-MITA).

42. *LL*, 2:185. Rogers took pride in finding that "some eminent scientific friends abroad including one of the directors of the Conservatoire des Arts et Metiers, and the ablest mathematical engineer of G. Britain expressed a very high appreciation of the scheme as set forth in the [*Scope and Plan*] pamphlet" (WBR to [?], January 1, 1865, WBRP-MITA).

43. WBR to HDR, *LL*, 2, April 13, 1864, *LL*, 2:191–92; Thomas Webb to WBR, June 7, 1864, *LL*, 2:192; WBR to HDR, July 23, 1864, *LL*, 2:199; WBR to RER, August 26, 1864, *LL*, 2:205.

44. Prescott, *Boston Tech*, 46–49.

45. *LL*, 2:224; Prescott, *Boston Tech*, 51.

46. Hugh Hawkings, *Between Harvard and America: The Educational Leadership of Charles W. Eliot* (New York: Oxford University Press, 1972); CWE to WBR, June 20, 1865, *LL*, 1:243; CWE to Arthur T. Lyman, April 18, 1865, reprinted in Henry James, *Charles W. Eliot: President of Harvard University, 1869–1909*, 2 vols. (New York: Houghton Mifflin, 1930), 1:143–47.

47. WBR to CWE, June 6, 1865, *LL*, 2:238–39; WBR to CWE, July 17, 1865, *LL*, 2:240–42; CWE to his mother, August 2, 1865, reprinted in James, *President of Harvard*, 1:155–56.

48. Prescott, *Boston Tech*, 54–59.

49. Paul Venable Turner, in *Campus: An American Planning Tradition* (Cambridge, Mass.:

MIT Press, 1984), has also called attention to this contrast between the exterior and interior of what came to be known as the Rogers Building. Turner interprets the contrast as reflecting an "ambivalence about whether grandeur or stark utilitarianism was the proper image for a technical school" (164). Rather than the result of "ambivalence," the building presents a reflection of Rogers's useful arts educational plan.

50. *LL,* 2:281.

Chapter Eight · Reception of the Idea

1. *LL,* 2:181; JDR to Emma Savage, February 9, 1869, *LL,* 2:287; WBR to HDR, November 17, 1863, *LL,* 2:180; *LL,* 2:304, 306.

2. *LL,* 2:279, 284, 302; Samuel Kneeland to WBR, December 3, 1864, *LL,* 2:285; WBR to the Government of the Massachusetts Institute of Technology, May 3, 1870, *LL,* 2:295; Government of the Institute to WBR, May 20, 1870, *LL,* 2:296–97.

3. "Technical Education," *Scientific American,* April 18, 1868, 249. Religion, by this point, would not have entered significantly into the debate. See Jon H. Roberts and James Turner, *The Sacred and the Secular University* (Princeton: Princeton University Press, 2000), who have argued that during this period "most scientists, Christian and otherwise, no longer judged the effectiveness of their efforts by whether they enable human beings to 'satisfy the aspirations of Reason to understand the wisdom of the Creator in his work.' In fact, religious concerns became essentially extrinsic to the culture of science" (31).

4. While scholarship by Paul Mattingly and David Potts has complicated the traditional view of the Yale Report of 1828, the use of the report and allusions to it in the postbellum period are the focus of this analysis. Noah Porter, *The American College and the American Public* (New Haven, Conn.: C. C. Chatfield, 1870), 42, 92, 46–48, 93, 154–55.

5. Porter, *American College,* 271. Porter was not simple-mindedly opposed to the introduction of modern studies in the curriculum. For studies that suggest that he and the circle of New Haven scholars appreciated German ideals of research and inquiry, see Louise L. Stevenson, *Scholarly Means to Evangelical Ends: The New Haven Scholars and the Transformation of Higher Learning in America, 1830–1890* (Baltimore: Johns Hopkins University Press, 1986); George M. Marsden, *The Soul of the American University: From Protestant Establishment to Established Non-Belief* (New York: Oxford University Press, 1994).

6. [G. P. Fisher], "The Memoir of President Wayland," *New Englander* (January 1868): 70–71; [T. W. Higginson], "A Plea for Culture," *Atlantic Monthly* (January 1867): 30–32.

7. [J. Jackson Jarves], "Museums of Art, Artists, and Amateurs in America," *Galaxy* (July 1870): 55–56; [P. A. Chadbourne], "Colleges and College Education," *Putnam's Monthly* (September 1869): 336, 339, 340.

8. "Jacob Bigelow," *Dictionary of American Biography,* 22 vols. (New York: Charles Scribner's Sons, 1929), 2:257–58; *Proceedings, Massachusetts Historical Society* 17 (March 1880): 383–467; on Bigelow and the Rumford Professorship, see Mary Ann James, "Engineering an Environment of Change: Bigelow, Peirce, and Early Nineteenth-Century Practical Education at Harvard," in Clark A. Elliott and Margaret W. Rossiter, *Science at Harvard University: Historical Perspectives* (Bethlehem, Pa.: Lehigh University Press, 1992), 59–63; William P. Atkin-

son, *Dynamic and Mechanic Teaching: A Lecture Read before the American Institute of Instruction, at the Annual Meeting, in New Haven, Conn., August 9th, 1865* (Cambridge: Sever and Francis, 1866), 3.

9. Jacob Bigelow, *An Address on the Limits of Education, Read before the Massachusetts Institute of Technology, November 16, 1865* (Boston: E. P. Dutton, 1865), 6, 15.

10. Jacob Bigelow, "On Classical and Utilitarian Studies. Read before the American Academy of Arts and Sciences, December 20, 1866," reprinted in Bigelow, ed., *Modern Inquiries: Classical, Professional, and Miscellaneous* (Boston: Little, Brown, , 1867), 39, 79, 87.

11. Bigelow, "Classical and Utilitarian Studies," 60, 46.

12. William P. Atkinson, *Classical and Scientific Studies, and the Great Schools of England: A Lecture Read before the Society of Arts of the Massachusetts Institute of Technology, April 6, 1865* (Cambridge: Sever and Francis, 1865), v, 17. "All evidence," claimed Atkinson, "goes to show that the English classical system is a portentous failure" (37).

13. Atkinson, *Schools of England,* 26, 28.

14. "Modern Inquiries," *New Englander* (January 1868): 210.

15. [J. E. Cabot], "Bigelow's Address on the Limits of Education," *North American Review* (April 1866): 592–93, 597–98; [Cabot], "Bigelow's Address," 596–97; [Cabot], "Bigelow's Classical and Utilitarian Studies," *North American Review* (April 1867): 612, 614–15; [J. Fiske], "Bigelow's Modern Inquiries," *North American Review* (July 1867): 298–301.

16. [J. E. Cabot], "Atkinson's Classical and Scientific Studies," *North American Review* (October 1865): 579, 581, 584.

17. WBR to E. L. Youmans, April 21, 1867, WBRP-MITA; E. L. Youmans to WBR, April 18, 1867, WBRP-MITA. Youmans's lecture was to be based on "Introduction—On Mental Discipline in Education," in E. L. Youmans, ed., *The Culture Demanded by Modern Life; A Series of Addresses and Arguments on the Claims of Scientific Education* (New York: D. Appleton and Co., 1867), 1–56.

18. WBR to E. L. Youmans, April 21, 1867, WBRP-MITA.

19. *Objects and Plan of an Institute of Technology* (Boston: J. Wilson, 1861), 28; *Scope and Plan of the School of Industrial Science, Massachusetts Institute of Technology* (Boston: J. Wilson and Son, 1864); *First Annual Catalogue of the Officers and Students and Programme of the Course of Instruction of the School of the Massachusetts Institute of Technology, 1865-6* (Boston: J. Wilson and Son, 1865). Rogers also relied on one other document to clarify the Institute's mission: *An Account of the Proceedings Preliminary to the Organizations of the Massachusetts Institute of Technology* (Boston: J. Wilson and Son, 1861).

20. WBR to Youmans, April 21, 1867; Rogers, *Objects and Plan,* 28.

21. "Technical Education," *Scientific American,* April 18, 1868, 249.

22. "Scientific Versus Classical Education," *Scientific American,* August 25, 1860, 137.

23. [P. H. Van Der Weyde, ed.], "Common Sense vs. Classical Education," *Manufacturer and Builder* (August 1872): 185.

24. [P. H. Van Der Weyde, ed.], "Mechanical Education," *Manufacturer and Builder* (June 1872): 138–39.

25. Edward Lurie, *Louis Agassiz: A Life in Science* (Baltimore: Johns Hopkins University Press, 1988), 362; Clark A. Elliot and Margaret W. Rossiter, eds., *Science at Harvard University: Historical Perspectives* (Bethlehem, Pa.: Lehigh University Press, 1992); Joseph LeConte,

"Morphology and Its Connection with Fine Art," *Southern Presbyterian Review* 12 (1859): 109. Julie Reuben, in *Making of the Modern University: Intellectual Transformation and the Marginalization of Morality* (Chicago: University of Chicago Press, 1996), has argued that "a fairly simple Baconian model dominated public discussions of science in the United States until the late nineteenth century" (38). Daniel Kevles makes a similar point in *The Physicists: The History of a Scientific Community in Modern America* (Cambridge: Harvard University Press, 1995): "The Baconian tradition no doubt encouraged American physicists to pursue an arid form of empiricism; the importance of facts to science was all too easily transmuted into an emphasis on mere fact-gathering" (37). To this we can add mid-nineteenth-century debates over applied, abstract, and useful arts approaches to scientific inquiry.

26. CWE, "The New Education," *Atlantic Monthly* 23 (1869): 206.

27. [Charles G. Leland], "Polytechnic Institutes," *Continental Monthly* (July 1862): 83–84, 86, 89; *Objects and Plan of an Institute of Technology* (Boston: J. Wilson, 1861).

28. Horace E. Scudder, "Education by Hand," *Harper's Monthly* (February 1879): 411. Unlike practitioners of law or medicine, engineers and others involved with the practical applications of science were not organized enough to counteract the replacement of "shop culture" with "school culture," or apprenticeships with institutes. Where practicing lawyers and physicians asserted influence over the shape of professional education, practitioners of the useful arts lacked a similar movement during most of the nineteenth century. See Michael Burrage's comparative study "From Practice to School-based Professional Education: Patterns of Conflict and Accommodation in England, France, and the United States," in Sheldon Rothblatt and Bjorn Wittrock, eds., *The European and American University since 1800: Historical and Sociological Essays* (New York: Cambridge University Press, 1993), 142–87.

29. Edward Atkinson, "Elementary Instruction in the Mechanic Arts," *Scribner's Monthly* (March 1881): 905–6; [William H. Rideing], "How to Become a Mechanical Engineer," *Scribner's Monthly* (May 1879): 141; "Literary," *Manufacturer and Builder* (December 1870): 369; [William H. Wahl, ed.], "Engineering at Cornell," *Manufacturer and Builder* (May 1882): 98.

30. Bruce Sinclair, "Harvard, MIT, and the Ideal Technical Education," in Elliott and Rossiter, eds., *Science at Harvard University*, 78.

31. Julius A. Stratton and Loretta Mannix, *Mind and Hand: The Birth of MIT* (Cambridge, Mass.: MIT Press, 2005), 508.

32. WBR to William P. Atkinson, June 8, 1865, WBRP-MITA; Albert J. Wright to WBR, October 9, 1865, WBRP-MITA; A. H. Russell to Emma Savage, May 28, 1881, WBRP-MITA; A Hyatt to WBR, October 3, 1878, WBRP-MITA.

33. Students to the MIT Corporation, January 2, 1880, WBRP-MITA; WBR notes on student interviews, January 2 and 5, 1880, WBRP-MITA.

34. On separate spheres and coeducation, see David Tyack and Elizabeth Hansot, *Learning Together: A History of Coeducation in American Public Schools* (New York: Russell Sage Foundation, 1992). Tyack and Hansot argue that "most public-school educators ignored or rebuffed Clarke, largely because their experience of coeducation did not substantiate his views and because they could see no practical way to answer his objections to the education of adolescent girls" (154); see also Janice Law Trecker, "Sex, Science and Education," *American Quarterly* 26 (1974): 352–66; Sue Zschoche, "Dr. Clarke Revisited: Science, True Womanhood, and Female Collegiate Education," *History of Education Quarterly* 29 (1989): 545–69; Leslie Miller-

Bernal, Separate by Degree: Women Students' Experiences in Single-Sex and Coeducational Colleges (New York: Peter Lang, 2000); Jane Hunter, *How Young Ladies Became Girls: The Victorian Origins of American Girlhood* (New Haven: Yale University Press, 2002).

35. On women and science education, see Barbara Miller Solomon, *In the Company of Educated Women: A History of Women and Higher Education in America* (New Haven: Yale University Press, 1985); Lois Barber Arnold, *Four Lives in Science: Women's Education in the Nineteenth Century* (New York: Schocken Books, 1984); Margaret W. Rossiter, *Women Scientists in America: Struggles and Strategies to 1940* (Baltimore: Johns Hopkins University Press, 1982); Deborah Jean Warner, "Science Education for Women in Antebellum America," *Isis* 69 (1978): 58–67; Kim Tolley, "Science for Ladies, Classics for Gentlemen: A Comparative Analysis of Scientific Subjects in the Curricula of Boys' and Girls' Secondary Schools in the United States, 1794–1850," *History of Education Quarterly* 36 (1996): 129–53. Further research by Tolley, *The Science Education of American Girls: A Historical Perspective* (New York: Routledge-Falmer, 2003), traces social, cultural, and other forces that influenced science and mathematics education for women from the middle of the eighteenth century to the middle of the twentieth century. For a case study of women's science education at the collegiate level, see Miriam R. Levin, *Defining Women's Scientific Enterprise: Mount Holyoke Faculty and the Rise of American Science* (Hanover, N.H.: University Press of New England, 2005). W. P. Atkinson to WBR, August 18, 1867, *LL*, 2:275–76; WBR to N[?]. Thayer, February 4, 1867, *LL*, 2:269.

36. Thomas T. Bouve to WBR, April 5, 1876, box 5, folder 72, WBRP-MITA.

37. Caroline Hunt, *The Life of Ellen H. Richards* (Boston: Whitcomb and Barrows, 1912), 88; Ellen Swallow Richards, "Report for the April 1882 Meeting of the Women's Education Association," folder 11, Records of the Women's Laboratory, 1875–1922, AC 298, MITA; Jessica Scalzi Ancker, "Domesticity, Science, and Social Control: Ellen Swallow Richards and the New England Kitchen" (B.A. thesis, Harvard University, 1987); Marilynn A. Bever, "The Women of MIT, 1871 to 1941: Who They Were, What They Achieved," 2 vols. (B.S. thesis, MIT, 1976); Ruth Schwarts Cowan, "Ellen Swallow Richards: Technology and Women," in Caroll W. Pursell Jr., ed., *Technology in America: A History of Individuals and Ideas* (Cambridge, Mass.: MIT Press, 1981), 142–50; JDR to WBR, July 24, 1876, *LL*, 2:336. Runkle's deference to Rogers's views on coeducation and other matters concerning the Institute is well documented in correspondence between the two. On major decisions affecting MIT, Runkle consistently deferred to Rogers whenever possible for the "final word." See JDR to CWE, February 2, 1870, cited in Prescott, *Boston Tech*, 76.

38. "Education," *Atlantic Monthly* (June 1877): 767–68.

39. Marian Hovey to WBR, December 13, 1878, WBRP-MITA.

40. Kneeland to WBR, February 19, 1876, WBRP-MITA.

41. WBR to Committee (notes), September 27, 1879, WBRP-MITA; Institute professor William P. Atkinson, Edward's brother, promoted coeducation and the interest of women at the Institute and hired female assistants; see, for example, William P. Atkinson to WBR, May 23, 1879, WBRP-MITA; William P. Atkinson to WBR, September 24, 1879, WBRP-MITA.

42. Edward Atkinson to WBR, October 17, 1879, box 6, folder 94, WBRP-MITA; WBR to Prof. Ordway, October 24, 1879, WBRP-MITA.

43. *First Annual Catalogue of the Officers and Students, and Programme of the Course of In-*

struction of the School of the Massachusetts Institute of Technology, 1865–6 (Boston: J. Wilson and Son, 1865), 10, 17, 20.

44. JDR to Emma Savage and WBR, July 5, 1876, WBRP-MITA.

45. For a discussion of the Russian tool system, see William John Schurter, "The Development of the Russian System of Tool Instruction (1763–1893) and Its Introduction into U.S. Industrial Education Programs (1876–1893)" (Ph.D. diss., University of Maryland, 1982). I am indebted to Schurter for calling my attention to relevant primary sources and historiography; JDR to Charles Ham, May 22, 1884, cited in Schurter, "Russian System," 183; JDR, *The Manual Element in Education. From the Forty-first Annual Report of the Board of Education, 1876–77* (Boston: Albert J. Wright, 1878), 188, 133.

46. JDR to WBR, July 5 and 24, 1876, *LL,* 2:335–36; *LL,* 2:337–38; JDR to Emma Savage and WBR, July 5, 1876, WBRP-MITA. Runkle's published descriptions of the Russian system include *The Russian System of Shop-work Instruction for Engineers and Machinists* (Boston: A. A. Kingman, 1876); "The Russian System of Shop-work Instruction," *President's Report for the Year Ending Sept. 30, 1876* (Boston: A. A. Kingman, 1876), 124–47.

47. [Charles Barnard], "New Roads to a Trade," *Century* (November 1881–April 1882): 287; "Editor's Table," *Appleton's Journal* (November 1876): 474–75.

48. Scudder, "Education by Hand," 414.

49. [P. H. Vander Weyde, ed.], "Teaching Practical Mechanics," *Manufacturer and Builder* (March 1877): 63; [P. H. Van Der Weyde, ed.], "Industrial Education in Boston," *Manufacturer and Builder* (April 1878): 81; [John A. Church], "Instruction Shops in Boston," *Galaxy* (March 1876): 422.

50. Eliot described his efforts as an attempt to create a "union" between Harvard and MIT. In reality, however, the union was not one that Harvard considered equal. Indeed, the proposals called for discarding the Lawrence School and replacing it with MIT, as a faculty under Harvard control. For other accounts about the attempts to "merge" Harvard and MIT in the 1870s, see Hector James Hughes, "Engineering," in Samuel Eliot Morison, ed., *The Development of Harvard University since the Inauguration of President Eliot, 1869–1929* (Cambridge: Harvard University Press, 1930), 413–42; Samuel C. Prescott, *When MIT Was "Boston Tech," 1861–1916* (Cambridge, Mass.: MIT Press, 1954), 71–87; and Sinclair, "Harvard, MIT, and the Ideal Technical Education," 22, 42–43, 75–79.

51. JDR to Emma Savage, April 8, 1869, *LL,* 2:287; CWE to WBR, July 12, 1869, WBRP-MITA; CWE, "Inaugural Address as President of Harvard College," in *Educational Reform: Essays and Addresses* (New York: Century Co., 1901), 6; see also Harvard University Archives Collection, Harvard-MIT Proposed Merger Papers and Harvard Miscellaneous Papers (1860–82).

52. JDR to WBR, December 23, 1869, WBRP-MITA; JDR to WBR, January 27, 1870, WBRP-MITA; WBR to JDR, February 1, 1870, *LL,* 2:293.

53. WBR to JDR, February 1, 1870, *LL,* 2:293; CWE to JDR, February 4, 1870, WBRP-MITA; Samuel Kneeland to WBR, February 14, 1870, WBRP-MITA; R. C. Greenleaf to WBR, July 28, 1870, WBRP-MITA; CWE to [James Lawrence?] March 23, 1871, Eliot Papers, Harvard University, cited in Sinclair, "Harvard, MIT, and the Ideal Technical Education," 84; Prescott, *Boston Tech,* 74–75.

54. JDR to CWE, February 2, 1870, cited in Prescott, *Boston Tech,* 76; WBR to CWE, February 7, 1870, *LL,* 2:294; WBR to JDR, February 22, 1870, WBRP-MITA.

55. CWE to WBR, February 9, 1870, cited in Prescott, *Boston Tech,* 77–78; WBR to JDR, February 22, 1870, *LL,* 2:294.

56. JDR to WBR and Emma Savage, June 22, 1870, WBRP-MITA; JDR to WBR, August 6, 1870, WBRP-MITA; Prescott, *Boston Tech,* 78–81. The idea of merger was also discussed in Edward Atkinson to WBR, July 28, 1870, WBRP-MITA; JDR to WBR, August 1, 1870, WBRP-MITA; JDR to WBR, August 2, 1870, WBRP-MITA; A. S. Wheeler to William Endicott, August 6, 1870, WBRP-MITA; Henry B. Rogers to WBR, July 20, 1870, WBRP-MITA; WBR to R. C. Greenlief, July 26, 1870, WBRP-MITA; JDR to WBR, July 27, 1870, WBRP-MITA.

57. JDR to WBR, September 9, 1870, folder 65, box 5, WBRP-MITA; see also Prescott, *Boston Tech,* 85–87; "Invitation to the Government of the Institute of Technology," *Annual Reports of the President and Treasurer of Harvard College, 1869–1870* (Cambridge: Cambridge University Press, 1871), 68; "Memorandum of an Agreement between Harvard College and the Mass. Inst. of Technology to Effect a Union of Their Several Schools of Applied Science," folder 1486, box 140, Records of President CWE, Harvard University Archives; correspondence on the merger proposal faded after JDR to WBR, September 5, 1870, WBRP-MITA; JDR to WBR, September 22, 1870, WBRP-MITA.

58. Survey Board of Washington College to Jedediah Hotchkiss, June 29, 1868, cited in Peter W. Roper, "Hotchkiss and the Geological Map of Virginia," *Earth Sciences History* 10 (1991): 40; see also Peter Lessing, "The Rogers-Hotchkiss Geological Maps of Virginia and West Virginia," *Earth Sciences History* 14 (1995): 84–97; Robert H. Silliman, "The Richmond Boulder Trains: Verae Causae in 19th-Century American Geology," *Earth Sciences History* 10 (1991): 60–72. In the postwar period Hotchkiss's training also led to temporary appointments with the Chesapeake and Ohio Railroad and the state's Board of Immigration. Each of these projects demanded a knowledge and presentation of the Virginia terrain, topography, and geology.

59. WBR's works for the period include "On the Gravel and Cobblestone Deposits of Virginia and the Middle [Atlantic] States," *Boston Society of Natural History Proceedings* 18 (1875): 101–6; *Hotchkiss' Geological Map of Virginia and West Virginia: The Geology by Prof. W. B. Rogers* (Richmond, Va.: A. Hoen and Co., 1875); "On the Newport Conglomerate [Rhode Island]." *Boston Society of Natural History Proceedings* 18 (1875): 97–101; "Catalogue of the Note Books of the Geological Survey of Virginia, 1835–1877, by Prof. W. B. Rogers and His Assistants" (MS, ca. 1878); "List of Geological Formations Found in Virginia and West Virginia," in James Macfarlane, ed., *Geologists' Travelling Hand-book, or American Geological Railway Guide* (New York: Appleton, 1879); "Table of the Geological Formations Found in Virginia and West Virginia," *Virginias: A Mining, Industrial, and Scientific Journal* 1 (1880).

60. Rexmond C. Cochrane, *The National Academy of Sciences: The First Hundred Years, 1863–1963* (Washington, D.C.: National Academy of Sciences, 1978), 118; WBR to Joseph Henry, November 14, 1872, *LL,* 2:307–8.

61. Simon Newcomb, *Reminiscences of an Astronomer* (Boston: Houghton, Mifflin, 1903), 250–51; Cochrane, *National Academy of Sciences,* 100–133.

62. Cochrane, *National Academy of Sciences,* 127–33; *LL,* 2:357–59; Thomas G. Manning, *Government in Science: The U.S. Geological Survey, 1867–1894* (Lexington: University of Kentucky Press, 1967), 30–59; O. C. Marsh to WBR, November 19, 1878, *LL,* 2:358.

63. Cochrane, *National Academy of Sciences,* 134–39; National Academy of Sciences, *A History of the First Half-Century of the National Academy of Sciences, 1863–1913* (Washington, D.C.: National Academy of Sciences, 1913), 46–55.

64. *LL,* 2:320, 338, 343; James D. Dana to WBR, August 17, 1877, *LL,* 2:343; WBR to James D. Dana, August 26, 1877, *LL,* 2:344; Mr. Bouve to WBR, April 17, 1880, *LL,* 2:365–66.

65. JDR to CWE, February 2, 1870, cited in Prescott, *Boston Tech,* 76.

66. *LL,* 2:350–51; E. R. Mudge to WBR, June 18, 1878, WBRP-MITA.

67. James P. Munroe, *A Life of Francis Amasa Walker* (New York: Henry Holt, 1923), 206; MIT Committee of the Faculty to WBR, May 23, 1881, *LL,* 2:371; *LL,* 2:377–82; WBR to Jedediah Hotchkiss, March 15, 1882, Hotchkiss Papers, Library of Congress, cited in Roper, "Geological Map," 41.

Chapter Nine · This Fatal Year

1. MIT Society of Arts, *In Memory of William Barton Rogers, LL.D., Late President of the Society* (Boston: Society of Arts, 1882); Emma Savage describes the last moments of WBR's life in *LL,* 2:387–389; Richmond *Daily Dispatch,* June 1, 1882; George F. Barker to Emma Savage, May 31, 1882, WBRP-MITA; Henry Bowditch to Emma Savage, May 31, 1882, WBRP-MITA; Spencer Baird to Emma Savage, June 4, 1882, WBRP-MITA; Thomas Wentworth Higginson to Emma Savage, June 5, 1882, WBRP-MITA; Wm Ripley Nichols to Emma Savage, June 1882, WBRP-MITA; J. S. Cabell to Emma Savage, June 10, 1882, WBRP-MITA; Henry B. Rogers to Emma Savage, August 7, 1882, WBRP-MITA; Asa Gray to John William Dawson, June 7, 1882 cited in Susan Sheets-Pyenson, *John William Dawson: Faith, Hope, and Science* (Montreal: McGill-Queen's University Press, 1996), 182.

2. MIT Society of Arts, *Rogers, LL.D.,* 20–21.

3. Henry A. Rowland, "A Plea for Pure Science," *Proceedings, American Association for the Advancement of Science* 32 (1883): 108, 111; Robert Julius Kwik, "The Function of Applied Science and the Mechanical Laboratory during the Period of Formation of the Profession of Mechanical Engineering, as Exemplified in the Career of Robert Henry Thurston, 1839–1903" (Ph.D. diss., University of Pennsylvania, 1974), 44.

4. CWE to WBR, July 12, 1869, WBRP-MITA; CWE, *Educational Reform: Essays and Addresses* (New York: Century Co., 1901), 318–19; Linda Chisholm "The Art of Undergraduate Teaching in the Age of the Emerging University" (Ph.D. diss., Columbia University, 1982), 201, 221, 279.

5. JDR to Emma Savage, June 12, 1869, WBRP-MITA; Louis Agassiz to Benjamin Peirce, October 26, 1868, Benjamin Peirce Papers, Houghton Library, Harvard University, Cambridge; John LeConte to WBR, October 16, 1876, WBRP-MITA. In 1864 an effort to establish an "Institute of Technology" in New York borrowed Rogers's *Objects and Plan* wholesale. Rogers wrote a letter to the editor of the New York *Evening Post* to call attention to the similarities and to suggest that MIT "will rejoice to welcome a sister Institute in New York, and cannot but be gratified at the reproduction in your city in such unchanged form of an educational plan in many respects new, and which we feel proud to have originated" (*New York Evening Post,* April 6, 1864). See also "A Proposed Institute of Technology," *New York Evening Post,* April 1, 1864. MIT's influence on laboratory work at Harvard is discussed in Lawrence

Aronovitch, "The Spirit of Investigation: Physics at Harvard University, 1870–1910," in Frank A.J.L. James, ed., *The Development of the Laboratory: Essays on the Place of Experiment in Industrial Civilization* (New York: American Institute of Physics, 1989), 83–103.

6. For the Walker-Shaler debate as well as other writings around and about that time, see Francis A. Walker, "Immediate Problems in Technological Education [Address Delivered at the International Congress of Education, Chicago, July 26, 1893]" reprinted in James Phinney Munroe, *Discussions in Education by Francis A. Walker, Ph.D., LL.D.* (New York: Henry Holt, 1899), 4, 12; Francis A. Walker, "Technological and Technical Education [Address Delivered at the Clarkson Memorial School of Technology, Potsdam, New York, November 30, 1896]," reprinted in Munroe, *Discussions in Education,* 96; Nathaniel Southgate Shaler, "Relations of Academic and Technical Instruction," *Atlantic Monthly* 72 (August 1893): 261, 263; N. S. Shaler to Horace Scudder, undated, cited in David N. Livingstone, *Nathaniel Southgate Shaler and the Culture of American Science* (Tuscaloosa: University of Alabama Press, 1987), 270; Francis A. Walker to Horace Scudder, May 17, 1893, cited in Livingstone, *Nathaniel Southgate Shaler,* 270; Francis A. Walker, "The Technical School and the University," *Atlantic Monthly* 72 (September 1893): 390–92. James Phinney Munroe, *A Life of Francis Amasa Walker* (New York: Henry Holt, 1923), 211–34; Bruce Sinclair, "Harvard, MIT, and the Ideal of Technical Education," in Clark A. Elliot and Margaret W. Rossiter, eds., *Science at Harvard University: Historical Perspectives* (Bethlehem, Pa.: Lehigh University Press, 1992); Samuel Eliot Morison, *Three Centuries of Harvard, 1636–1936* (Cambridge: Harvard University Press, 1936), 371–72, 471.

7. James Phinney Munroe, *A Life of Francis Amasa Walker* (New York: Henry Holt, 1923), 211–34; Bruce Sinclair, "Harvard, MIT, and the Ideal of Technical Education," in Clark A. Elliot and Margaret W. Rossiter, eds., *Science at Harvard University: Historical Perspectives* (Bethlehem, Pa.: Lehigh University Press, 1992); Samuel Eliot Morison, *Three Centuries of Harvard, 1636–1936* (Cambridge: Harvard University Press, 1936), 371–72, 471.

8. Emma Savage, Preface, in *GV,* iii; Robert Rakes Shrock, *Geology at MIT, 1865–1965: A History of the First Hundred Years of Geology at the Massachusetts Institute of Technology,* 2 vols. (Cambridge, Mass.: MIT Press, 1977–82), 2:437–39.

MANUSCRIPT COLLECTIONS

College of William and Mary Archives, Williamsburg, Va.
    College Papers: 1819–35
    Faculty Minutes: 1819–35
    Patrick Kerr Rogers, Faculty and Alumni Papers
    William Barton Rogers, Faculty and Alumni Papers
Harvard University Archives, Cambridge
    Annual Reports of the President and Treasurer
    Harvard-MIT Proposed Merger Papers
    Harvard in 1860–82, Miscellaneous Papers
    Records of President Charles W. Eliot
Houghton Library, Harvard University, Cambridge
    A. D. Bache Papers
    Benjamin Peirce Papers
Library of Virginia Archives, Richmond
    Board of Public Works Collection
    Geological Survey Papers
Massachusetts Institute of Technology Archives (MIT), Cambridge
    Course Catalogue
    Faculty and Corporation Minutes
    John Daniel Runkle Papers
    Rogers Family Papers
    William Barton Rogers Papers
Massachusetts State Archives, Boston
    Acts and Resolves of Massachusetts
    Documents Printed by Order of the House of Representatives
    Documents Printed by Order of the Senate
National Academy of Sciences Archives (NAS), Washington, D.C.
    A. D. Bache, Member File
    Benjamin Peirce, Member File

Joseph Henry, Member File
Robert Empie Rogers, Member File
William Barton Rogers, Member File
University of Virginia Archives, Charlottesville, Va.
Faculty and Board of Visitors Minutes: 1835–53
University Catalogue: 183553
University Papers: 1835–53
William Barton Rogers Papers

NEWSPAPERS AND SERIAL PUBLICATIONS

Appleton's Journal
Atlantic Monthly
Century
Continental Monthly
Galaxy
Harper's Monthly
Manufacturer and Builder
New Englander
New York Evening Post
North American Review
Phoenix Plough-Boy (Williamsburg, Va.)
Putnam's Monthly
Richmond Family Visitor
Scientific Monthly
Scribner's Monthly
Southern Presbyterian Review

PUBLISHED DOCUMENTS

An Account of the Proceedings Preliminary to the Organizations of the Massachusetts Institute of Technology. Boston: J. Wilson and Son, 1861.

Agassiz, Elizabeth Cary, ed. *Louis Agassiz: His Life and Correspondence.* New York: Houghton Mifflin, 1885.

Anderson, Martin Brewer. *Sketch of the Life of Professor Chester Dewey, D.D., LL.D., Late Professor of Chemistry and Natural History in the University of Rochester.* Albany, 1868.

Atkinson, William P. *Classical and Scientific Studies, and the Great Schools of England: A Lecture Read before the Society of Arts of the Massachusetts Institute of Technology, April 6, 1865.* Cambridge: Sever and Francis, 1865.

———. *Dynamic and Mechanic Teaching: A Lecture Read before the American Institute of Instruction, at the Annual Meeting, in New Haven, Conn., August 9th, 1865.* Cambridge: Sever and Francis, 1866.

Bache, A. D., and Henry Darwin Rogers. "Analysis of Some Coals of Pennsylvania." *Journal of the Academy of Natural Sciences of Philadelphia* 7 (1834): 158–77.

Baker, Ray Palmer. "Rensselaer Polytechnic Institute and the Beginnings of Science in the United States." *Scientific Monthly* 19 (October 1924): 337–56.

Barker, George Frederick. *Memoir of John William Draper, 1811–1882. By George F. Barker. Read before the National Academy, April 21, 1886.* Washington, D.C.: n.p., 1886.

Beckwith, Florence. *Early Botanists of Rochester and Vicinity and the Botanical Section.* Rochester: Rochester Academy of Science, 1912.

Bigelow, Jacob. *An Address on the Limits of Education, Read before the Massachusetts Institute of Technology, November 16, 1865.* Boston: E. P. Dutton, 1865.

———. *Elements of Technology, Taken Chiefly from a Course of Lectures Delivered at Cambridge, on the Application of the Sciences to the Useful Arts.* Boston: Hilliard, Gray, Little and Wilkins, 1831.

———. *Modern Inquiries: Classical, Professional, and Miscellaneous.* Boston: Little, Brown, 1867.

Boutwell, George S. *Thoughts on Educational Topics and Institutions.* Boston: Phillips, Sampson, 1859.

Brande, William Thomas. *A Manual of Chemistry.* London: J. Murray, 1819.

Brenaman, J. N. *A History of Virginia Conventions.* Richmond: J. L. Hill Printing Co., 1902.

Bush, George Gary, ed. *History of Higher Education in Massachusetts.* Washington, D.C.: GPO, 1891, 280–319.

Cooper, Lane. *Louis Agassiz as Teacher: Illustrative Extracts on His Method of Instruction.* Ithaca: Comstock Publishing Co., 1917.

Davis, John A. G. *An Exposition of the Proceedings of the Faculty of the University of Virginia in Relation to the Recent Disturbances at That Institution.* Charlottesville: J. Alexander, 1836.

Donovan, Edward. *Instructions for Collecting and Preserving Various Subjects of Natural History.* London: n.p., 1794.

Eliot, Charles W. "The New Education." *Atlantic Monthly* 23 (1869): 203–20.

———. *Educational Reform: Essays and Addresses.* New York: Century Co., 1901.

For the Establishment of a School of Arts. Memorial of the Franklin Institute, of the State of Pennsylvania, for the Promotion of the Mechanic Arts, to the Legislature of Pennsylvania. Philadelphia: J. Crissy, 1837.

Graves, George. *Naturalist's Pocket-Book and Tourist's Companion.* London: Longman, 1818.

Greene, Benjamin Franklin. *The Rensselaer Polytechnic Institute: Its Reorganization in 1849–50; Its Condition at the Present Time: Its Plans and Hopes for the Future.* Troy, N.Y.: D. H. Jones, 1855.

Hillard, George Stillman. *Memoir of the Hon. James Savage, LL.D., Late President of the Massachusetts Historical Society.* Boston: John Wilson and Son, 1878.

Hotchkiss' Geological Map of Virginia and West Virginia: The Geology by Prof. W. B. Rogers. Richmond: A. Hoen and Co., 1875.

"The Inequality of Representation in the General Assembly of Virginia: A Memorial to the Legislature of the Commonwealth of Virginia, Adopted at Full Meeting of the Citizens of Kanawha." *West Virginia History* 25 (July 1964): 283–98.

James, Henry. *Charles W. Eliot: President of Harvard University, 1869-1909.* 2 vols. New York: Houghton Mifflin, 1930.

Journal of the House of Delegates of the Commonwealth of Virginia . . . 1835. Richmond: Samuel Shepherd, 1835.

Journal of the House of Delegates of Virginia, Session 1841–42. Richmond: Samuel Shepard, 1841.

Journal of the House of Delegates of Virginia. Session 1844–45. Richmond: Samuel Shepherd, 1844.

Laws and Regulations of the College of William and Mary in Virginia. Richmond: Thomas W. White, 1830.

Lettsome, John. *Naturalist's and Traveller's Companion.* London: C. Dilly, 1774.

Macfarlane, James, ed. *Geologists' Travelling Hand-book, or American Geological Railway Guide.* New York: Appleton, 1879.

Marcou, Jules. *Life, Letters, and Works of Louis Agassiz.* New York: Macmillan, 1896.

Meigs, Charles D. *A Memoir of Samuel George Morton, M.D., Late President of the Academy of Natural Sciences of Philadelphia.* Philadelphia: T. K. and P. G. Collins, 1851.

MIT Society of Arts, *In Memory of William Barton Rogers, LL.D., Late President of the Society.* Boston: Society of Arts, 1882.

Munroe, James P. "The Beginning of the Massachusetts Institute of Technology." *Technology Quarterly* 1 (May 1888): 285–97.

Newcomb, Simon. *Reminiscences of an Astronomer.* Boston: Houghton, Mifflin, 1903.

Objects and Plan of an Institute of Technology. Boston: J. Wilson, 1861.

"Original Papers in Relation to a Course of Liberal Education." *American Journal of Science and Arts* 25 (1829): 197–351.

Patterson, Henry S. *Memoir of the Life and Scientific Labors of Samuel George Morton.* Philadelphia: Lippincott, Grambo and Co., 1854.

Picture of Baltimore, Containing a Description of All Objects of Interest in the City and Embellished with Views of the Principal Public Buildings. Baltimore: F. Lucas Jr., 1832.

Porter, Noah. *The American College and the American Public.* New Haven: C. C. Chatfield, 1870.

A Provisional List of Alumni, Grammar School Students, Members of the Faculty and Members of the Board of Visitors of the College of William and Mary in Virginia from 1693 to 1888. Richmond: Division of Purchase and Printing, 1941.

Public Documents of Massachusetts: Being the Annual Reports of Various Public Offices and Institutions for the Year 1864, vol. 1: *Public Document Number 4, Annual Report of the Massachusetts Board of Agriculture.* Boston: William White, 1865.

Pullman, David L. *The Constitutional Conventions of Virginia from the Foundation of the Commonwealth to the Present Time.* Richmond: J. T. West, 1901.

Quint, Alonzo. *The Potomac and the Rapidian.* Boston: Crosby and Nichols, 1864.

———. *The Record of the Second Massachusetts Infantry, 1861–65.* Boston: J. P. Walker, 1867.

Reports of the First, Second, and Third Meetings of the Association of American Geologists and Naturalists. Boston: Gould, Kendall and Lincoln, 1843.

A Reprint of Annual Reports and Other Papers on the Geology of the Virginias. New York: D. Appleton, 1884.

Rogers, Emma, ed. *Letters of James Savage to His Family.* Boston: n.p., 1906.

———. *Life and Letters of William Barton Rogers.* 2 vols. Boston: Houghton Mifflin, 1898.

Rogers, Patrick Kerr. *An Introduction to the Mathematical Principles of Natural Philosophy.* Richmond, Va.: Shepherd and Pollard, 1822.

Rogers, William Barton. "An Account of Apparatus and Processes for Chemical and Photometrical Testing of Illuminating Gas." *British Association Report* 34 (1864): 39–40.

———. *Address before the Lyceum of Natural History of Williams College, August 14, 1855.* Boston: T. R. Marvin and Son, 1855.

——— . "Catalogue of the Note Books of the Geological Survey of Virginia, 1835–1877, by Prof. W. B. Rogers and His Assistants." MS ca. 1878.

———. "Discovery of Paleozoic Fossils in Eastern Massachusetts." *American Journal of Science* 22 (1856): 296–98.

———. "Electrical Illumination at Boston Photometrically Measured." *American Journal of Science* 36 (1863): 307–8.

———. *An Elementary Treatise on the Strength of Materials.* Charlottesville: Tompkins and Noel, 1838.

———. *Elements of Mechanical Philosophy.* Boston: Thurston, Torrey, and Emerson, 1852.

———. "Literature [Review of *On the Origin of Species*]." [Boston] *Courier,* March 5, 1860.

———. "Observations and Queries Respecting Artesian Wells." *Farmers' Register* 7 (December 1834): 451–55.

———. "On the Formation of Rotating Rings by Air and Liquids under Certain Conditions of Discharge." *American Journal of Science* 26 (1858): 246–58.

———. "On the Gravel and Cobblestone Deposits of Virginia and the Middle States." *Boston Society of Natural History Proceedings* 18 (1875): 101–6.

———. "On the Group of Rocks Constituting the Base of the Paleozoic Series in the United States." *Proceedings of the Boston Society of Natural History* 7 (1861): 394–95.

———. "On the Newport Conglomerate." *Boston Society of Natural History Proceedings* 18 (1875): 97–101.

———. "On the Origin and Accumulation of the Protocarbonate of Iron in Coal Measures." *Proceedings of the Boston Society of Natural History* 5 (1856): 283–88.

———. "On the Transporting Power of Currents." *American Journal of Science* 5 (1848): 115–16.

———. "Report from the Committee of Schools and Colleges against the Expediency of Withdrawing the Fifteen Thousand Dollars Annuity from the University [Document No. 41]." *Journal of the House of Delegates. Session 1844–45.* Reprinted in Emma Rogers, ed. *Life and Letters of William Barton Rogers.* 2 vols. Boston: Houghton Mifflin, 1898. 1:399–412.

———. "Some Experiments on Sonorous Flames, with Remarks on the Primary Source of Their Vibrations." *American Journal of Science* 26 (1858): 1–15.

———. "Table of the Geological Formations Found in Virginia and West Virginia." *Virginias: A Mining, Industrial, and Scientific Journal* 1 (1880): 14–15.

Rogers, William Barton, C. T. Jackson, and J. H. Blake. "The Frozen Well of Brandon, Vermont." *Boston Society of Natural History Proceedings* 9 (1862): 72–81.

Rogers, William Barton, and Henry Darwin Rogers. "Experimental Enquiry into Some of the Laws of the Elementary Voltaic Battery." *American Journal of Science* 27 (1835): 39–61.

———. "Observations on the Geology of the Western Peninsula of Upper Canada, and the Western Part of Ohio." *Proceedings of the American Philosophical Society* 2 (1842): 120–25.

———. "On the Geological Age of the White Mountains." *American Journal of Science* 1 (1846): 411–21.

———. "On the Phenomena of the Great Earthquakes . . . to Elucidate Several Points in Geological Dynamics." *Proceedings of the American Philosophical Society* 3 (1843): 64–67.

———. "On the Physical Structure of the Appalachian Chain as Exemplifying the Laws Which Have Regulated the Elevation of Great Mountain Chains Generally." *Reports of the First, Second, and Third Meetings of the Association of American Geologists and Naturalists at Philadelphia in 1840 and 1841 and at Boston in 1842.* Boston: Gould, Kendall, and Lincoln, 1843. Reprinted in *A Reprint of Annual Reports and Other Papers on the Geology of the Virginias.* New York: D. Appleton, 1884, 599–642.

———. "Theory of Earthquake Action." *American Journal of Science* 45 (1843): 341–47.

Rogers, William Barton, and Robert Empie Rogers. "On the Absorption of Carbonic Acid Gas by Liquids." *American Journal of Science* 6 (1848): 96–110.

Ross, M. D. *Estimate of the Financial Effect of the Proposed Reservation of Back-Bay Lands.* Boston, J. Wilson and Son, 1861.

Rowland, Henry A. "A Plea for Pure Science." *Proceedings, American Association for the Advancement of Science* 32 (1883): 105–26.

Runkle, John D. *The Manual Element in Education. From the Forty-first Annual Report of the Board of Education, 1876–77.* Boston: Albert J. Wright, 1878.

———. *The Russian System of Shop-work Instruction for Engineers and Machinists.* Boston: A. A. Kingman, 1876.

Ruschenberger, W. S. W. *A Sketch of the Life of Robert E. Rogers, M.D., LL.D., with Biographical Notices of His Father and Brothers.* Philadelphia: McCalla and Stavely, 1885.

Scarborough, William K., ed. *The Diary of Edmund Ruffin: Toward Independence, October, 1856–April, 1861.* Baton Rouge: Louisiana State University Press, 1972.

Scope and Plan of the School of Industrial Science, Massachusetts Institute of Technology. Boston: J. Wilson and Son, 1864.

Shaler, Nathaniel Southgate. *The Autobiography of Nathaniel Southgate Shaler.* Boston: Houghton Mifflin, 1909.

———. "Relations of Academic and Technical Instruction." *Atlantic Monthly* 72 (August 1893): 259–68.

Smith, Charles C. "Memoir of John Amory Lowell, L. L. D." *Massachusetts Historical Society, Proceedings* (January 1898): 118–19.

Steiner, Bernard C. *The History of University Education in Maryland, Johns Hopkins University Studies in Historical and Political Science.* Baltimore: Johns Hopkins University Press, 1891.

Walker, Francis A. "The Technical School and the University." *Atlantic Monthly* 72 (September 1893): 390–95.

Warren, S. Edward. *Notes on Polytechnic or Scientific Schools in the United States: Their Nature, Position, Aims and Wants.* New York: Wiley, 1866.

Wayland, Francis. *A Memoir of the Life and Labors of Francis Wayland, D.D., LL.D., Late President of Brown University.* New York: Sheldon, 1867.

———. *Thoughts on the Present Collegiate System in the United States.* Boston: Gould, Kendall and Lincoln, 1842.

Webster, John White. *A Manual of Chemistry on the Basis of Professor Brande's*. Boston: Richard and Lord, 1828.

Wood, George B. *A Biographical Memoir of Samuel George Morton*. Philadelphia: T. K. and P. G. Collins, 1853.

Youmans, E. L., ed. *The Culture Demanded by Modern Life; A Series of Addresses and Arguments on the Claims of Scientific Education*. New York: D. Appleton and Co., 1867.

SECONDARY SOURCES

Adams, Sean Patrick. "Old Dominions and Industrial Commonwealths: The Political Economy of Coal in Virginia and Pennsylvania, 1810–1875." Ph.D diss., University of Wisconsin, Madison, 1999.

———. "Partners in Geology, Brothers in Frustration: The Antebellum Geological Surveys of Virginia and Pennsylvania." *Virginia Magazine of History and Biography* 106 (Winter 1998): 5–34.

Addis, Cameron. *Jefferson's Vision for Education, 1760–1845*. New York: Peter Lang, 2003.

Allmendinger, David F., Jr. *Ruffin: Family and Reform in the Old South*. New York: Oxford University Press, 1990.

Ambrose, Stephen. *Duty, Honor, Country: A History of West Point*. Baltimore: Johns Hopkins University Press, 1999.

Ancker, Jessica Scalzi. "Domesticity, Science, and Social Control: Ellen Swallow Richards and the New England Kitchen." B.A. thesis, Harvard University, 1987.

Archibald, Raymond Clare. "Unpublished Letters of James Joseph Sylvester and Other New Information concerning His Life and Work." *Osiris* 1 (January 1936): 85–154.

Arnold, Lois Barber. *Four Lives in Science: Women's Education in the Nineteenth Century*. New York: Schocken Books, 1984.

Artz, Frederick B. *The Development of Technical Education in France, 1500–1850*. Cleveland: Society for the History of Technology, 1966.

Axtell, James. "The Death of the Liberal Arts College." *History of Education Quarterly* 9 (1971): 339–52.

Ayers, Edward L. *Vengeance and Justice: Crime and Punishment in the 19th Century American South*. New York: Oxford University Press, 1984.

Beach, Mark. "Was There a Scientific Lazzaroni?" In George H. Daniels, ed. *Nineteenth Century American Science: A Reappraisal*. Evanston: Northwestern University Press, 1972.

Bell, Ian F. A. "Divine Patterns: Louis Agassiz and American Men of Letters, Some Preliminary Explorations." *Journal of American Studies* 10 (1976): 349–81.

Berkeley, Edmund, and Dorothy Smith Berkeley. *George William Featherstonhaugh: The First U.S. Government Geologist*. Tuscaloosa: University of Alabama Press, 1988.

Bever, Marilynn A. "The Women of MIT, 1871 to 1941: Who They Were, What They Achieved." 2 vols. B.S. thesis, MIT, 1976.

Binger, Carl Alfred Lanning. *Revolutionary Doctor: Benjamin Rush, 1746–1813*. New York: W. W. Norton, 1966.

Bonner, Thomas Neville. *Becoming a Physician: Medical Education in Britain, France, Germany, and the United States, 1750–1945.* New York: Oxford University Press, 1995.

Brain, Robert. *Going to the Fair: Readings in the Culture of Nineteenth Century Exhibitions.* Cambridge: Whipple Museum of the History of Science, 1993.

Brown, Chandos Michael. *Benjamin Silliman: A Life in the Young Republic.* Princeton: Princeton University Press, 1989.

Browning, William. "The Relation of Physicians to Early American Geology." *Annals of Medical History* 3 (1931): 547–60, 565.

Bruce, Dickson D., Jr. *The Rhetoric of Conservatism: The Virginia Convention of 1829–30 and the Conservative Tradition in the South.* San Marino, Calif.: Huntington Library, 1982.

———. *Violence and Culture in the Antebellum South.* Austin: University of Texas Press, 1979.

Bruce, Philip A. *History of the University of Virginia, 1819–1919: The Lengthened Shadow of One Man.* 5 vols. New York: Macmillan, 1920.

Bruce, Robert V. *Lincoln and the Tools of War.* Indianapolis: Bobbs-Merrill, 1956.

———. *The Launching of Modern American Science, 1846–1876.* New York: Knopf, 1987.

Burke, Colin Bradley. *American Collegiate Populations: A Test of the Traditional View.* New York: New York University Press, 1982.

Cary, Harold Whiting. *The University of Massachusetts: A History of One Hundred Years.* Amherst: University of Massachusetts Press, 1962.

Cash, W. J. *The Mind of the South.* New York: Random House, 1941.

Cassedy, James H. *Medicine in America: A Short History.* Baltimore: Johns Hopkins University Press, 1991.

Chisholm, Linda Armstrong. "The Art of Undergraduate Teaching in the Age of the Emerging University." Ph.D. diss., Columbia University, 1982.

Chute, William J. *Damn Yankee! The First Career of Frederick A. P. Barnard, Educator, Scientist, Idealist.* Port Washington, N.Y.: Kennikat Press, 1978.

Cochrane, Rexmond C. *The National Academy of Sciences: The First Hundred Years, 1863–1963.* Washington, D.C.: National Academy of Sciences, 1978.

Cohen, Benjamin R. "Surveying Nature: Environmental Dimensions of Virginia's First Scientific Survey, 1835–1842." *Environmental History* 11 (2006): 37–69.

Cooper, David Y., and Marshall A. Ledger. *Innovation and Tradition at the University of Pennsylvania School of Medicine: An Anecdotal Journey.* Philadelphia: University of Pennsylvania Press, 1990.

Cooper, William J., and Thomas E. Terrill. *The American South: A History.* New York: McGraw-Hill, 1996.

Corgan, James X., ed. *The Geological Sciences in the Antebellum South.* Tuscaloosa: University of Alabama Press, 1982.

Coulter, E. Merton. *College Life in the Old South.* Athens: University of Georgia Press, 1951.

Craven, Avery O. *Edmund Ruffin, Southerner: A Study in Secession.* New York: D. Appleton and Co., 1932.

———. *Soil Exhaustion as a Factor in the Agricultural History of Virginia and Maryland, 1606–1860.* Urbana: University of Illinois Press, 1926.

Croce, Paul Jerome. "Probabilistic Darwinism: Louis Agassiz vs. Asa Gray on Science, Religion, and Certainty." *Journal of Religious History* 22 (1998): 35–58.

Dabney, Virginius. *Mr. Jefferson's University: A History.* Charlottesville: University Press of Virginia, 1981.

Daniels, George H. *American Science in the Age of Jackson.* New York: Columbia University Press, 1968.

———. *Darwinism Comes to America.* Waltham: Blaisdell Publishing Co., 1968.

Davis, Tenney L. "Eliot and Storer: Pioneers in the Laboratory Teaching of Chemistry." *Journal of Chemical Education* 6 (1929): 868–79.

Day, Charles R. *Education for the Industrial World: The Ecole d'Arts et Metiers and the Rise of French Industrial Engineering.* Cambridge, Mass.: MIT Press, 1987.

D'Elia, Donald J. *Benjamin Rush: Philosopher of the American Revolution.* Philadelphia: American Philosophical Society, 1974.

Dexter, Ralph W. "The Impact of Evolutionary Theories on the Salem Group of Agassiz Zoologists (Morse, Hyatt, Packard, Putnam)." *Essex Institute Historical Collections* 115 (1979): 144–71.

Dictionary of American Biography. 22 vols. New York: Charles Scribner's Sons, 1928–58.

Dill, David B., Jr. "William Phipps Blake: Yankee Gentleman and Pioneer Geologist of the Far West." *Journal of Arizona History* 32 (1991): 385–412.

Dobbs, David. *Reef Madness: Alexander Agassiz, Charles Darwin, and the Meaning of Coral.* New York: Pantheon, 2005.

Dupree, A. Hunter. *Asa Gray: American Botanist, Friend of Charles Darwin.* Baltimore: Johns Hopkins University Press, 1988.

———. *Science in the Federal Government: A History of Policies and Activities.* Baltimore: Johns Hopkins University Press, 1986.

Dupuy, R. Ernest. *Sylvanus Thayer: Father of Technology in the United States.* New York: United States Military Academy, 1958.

Eaton, Clement. *The Freedom-of-Thought Struggle in the Old South.* New York: Harper and Row, 1964.

———. *The Mind of the Old South.* Baton Rouge: Louisiana State University Press, 1976.

Elliot, Clark A., and Margaret W. Rossiter, eds. *Science at Harvard University: Historical Perspectives.* Bethlehem: Lehigh University Press, 1992.

Ernst, William. "William Barton Rogers: Antebellum Virginia Geologist." *Virginia Cavalcade* 24 (Summer 1974): 13–20.

Ewing, G. W. *Early Teaching of Science at the College of William and Mary in Virginia.* Williamsburg, Va.: n.p., 1938.

Faler, Paul G. *Mechanics and Manufacturers in the Early Industrial Revolution: Lynn, Massachusetts, 1780–1860.* Albany: State University of New York Press, 1981.

Farrell, Betty. *Elite Families: Class and Power in Nineteenth-Century Boston.* Albany: State University of New York Press, 1993.

Faust, Drew Gilpin. *A Sacred Circle: The Dilemma of the Intellectual in the Old South, 1840–1860.* Baltimore: Johns Hopkins University Press, 1977.

Feuer, Lewis S. "America's First Jewish Professor: James Joseph Sylvester at the University of Virginia." *American Jewish Archives* 36 (November 1984): 152–201.

———. "Sylvester in Virginia." *Mathematical Intelligencer* 9 (1987): 3–19.

Finkelstein, Martin. "From Tutor to Specialized Scholar: Academic Professionalization in

Eighteenth and Nineteenth Century America." *History of Higher Education Annual* 3 (1983): 99–121.

Fleming, Donald. *John William Draper and the Religion of Science.* Philadelphia: University of Pennsylvania Press, 1950.

Forman, Sidney. *West Point: A History of the United States Military Academy.* New York: Columbia University Press, 1950.

Freehling, Alison Goodyear. *Drift toward Dissolution: The Virginia Slavery Debate of 1831–32.* Baton Rouge: Louisiana State University Press, 1982.

Freidel, Frank Burt. *Francis Lieber: Nineteenth-Century Liberal.* Baton Rouge: Louisiana State University Press, 1948.

Friend, Craig Thompson, and Lorri Clover, eds. *Southern Manhood: Perspectives on Masculinity in the Old South.* Athens: University of Georgia Press, 2004.

Furner, Mary O. *Advocacy and Objectivity: A Crisis in the Professionalization of American Social Science, 1865–1905.* Lexington: University Press of Kentucky, 1975.

Geer, William M. *Francis Lieber at the South Carolina College.* Raleigh: Print. Shop, North Carolina State College, 1943.

Geiger, Roger, ed. *The American College in the Nineteenth Century.* Nashville: Vanderbilt University Press, 2000.

Gerstner, Patsy A. "Henry Darwin Rogers and William Barton Rogers on the Nomenclature of the American Paleozoic Rocks." In Cecil J. Schneer, ed. *Two Hundred Years of Geology in America.* Hanover, N.H.: University Press of New England, 1979, 175–86.

———. "A Dynamic Theory of Mountain Building: Henry Darwin Rogers, 1842." *Isis* 66 (1975): 26–37.

———. *Henry Darwin Rogers, 1808–1866: American Geologist.* Tuscaloosa: University of Alabama Press, 1994.

Glick, Thomas F., ed. *The Comparative Reception of Darwinism.* Austin: University of Texas Press, 1974.

Godson, Susan H., et al. *The College of William and Mary: A History, 1693–1888.* 2 vols. Williamsburg, Va.: King and Queen Press, 1993.

Gohau, Gabriel. *A History of Geology.* New Brunswick, N.J.: Rutgers University Press, 1991.

Goldman, Lawrence. "Exceptionalism and Internationalism: The Origins of American Social Science Reconsidered." *Journal of Historical Sociology* (UK) 11 (March 1998): 1–36.

Goodman, Paul. "Ethics and Enterprise: The Values of a Boston Elite, 1800–1860." *American Quarterly* 18 (Fall 1966): 437–51.

Gorn, Elliott J. "'Gouge and Bite, Pull Hair and Scratch': The Social Significance of Fighting in the Southern Backcountry." *American Historical Review* 90 (February 1985): 18–43.

Greene, John C. *American Science in the Age of Jefferson.* Ames: Iowa State University Press, 1984.

Greene, Mott T. *Geology in the Nineteenth Century: Changing Views of a Changing World.* Ithaca: Cornell University Press, 1982.

Grimsted, David. *American Mobbing, 1828–1861.* New York: Oxford University Press, 1998.

Guralnick, Stanley M. *Science and the Antebellum American College.* Philadelphia: American Philosophical Society, 1975.

Haskell, Thomas L. *The Emergence of Professional Social Science: The American Social Science*

Association and the Nineteenth-Century Crisis of Authority. Urbana: University of Illinois Press, 1977.

Hawke, David Freeman. *Benjamin Rush: Revolutionary Gadfly.* Indianapolis: Bobbs-Merrill, 1971.

Hawkings, Hugh. *Between Harvard and America: The Educational Leadership of Charles W. Eliot.* New York: Oxford University Press, 1972.

Hermann, Kenneth W. "Shrinking from the Brink: Asa Gray and the Challenge of Darwinism, 1853–1868." Ph.D. diss., Kent State University, 1999.

Hoeveler, J. David., Jr. *James McCosh and the Scottish Intellectual Tradition: From Glasgow to Princeton.* Princeton: Princeton University Press, 1981.

Hofstadter, Richard, and Wilson Smith, eds. *American Higher Education: A Documentary History.* Chicago: University of Chicago Press, 1961.

Holden, Edward Singleton. *Biographical Memoir of William H. C. Bartlett, 1804–1893.* Washington, D.C.: National Academy of Sciences, 1911.

Holmfeld, John D. "From Amateurs to Professionals in American Science: The Controversy over the Proceedings of an 1853 Scientific Meeting." *Proceedings of the American Philosophical Society* 114 (February 1970): 22–36.

Hornberger, Theodore. *Scientific Thought in the American Colleges, 1639–1800.* Austin: University of Texas Press, 1945.

Hull, David L. *Darwin and His Critics: The Reception of Darwin's Theory of Evolution by the Scientific Community.* Cambridge: Harvard University Press, 1973.

Hunt, Caroline. *The Life of Ellen H. Richards.* Boston: Whitcomb and Barrows, 1912.

Hunter, Jane. *How Young Ladies Became Girls: The Victorian Origins of American Girlhood.* New Haven: Yale University Press, 2002.

Hurbst, Jurgen. "American Higher Education in the Age of the College." *History of Universities* 7 (1988): 37–59.

Jackson, James R., and William C. Kimler, "Taxonomy and the Personal Equation: The Historical Fates of Charles Girard and Louis Agassiz." *Journal of the History of Biology* (Netherlands) 32 (1999): 509–55.

James, Frank A. J. L., ed. *The Development of the Laboratory: Essays on the Place of Experiment in Industrial Civilization.* New York: American Institute of Physics, 1989.

James, I. M. "James Joseph Sylvester, F. R. S. (1814–1897)." *Notes and Records of the Royal Society of London* 51 (July 1997): 247–61.

Jardine, Nicholas, James A. Secord, and Emma C. Spary, eds. *Cultures of Natural History.* New York: Cambridge University Press, 1996.

Jefferson, Thomas. *Notes on the State of Virginia.* Chapel Hill: University of North Carolina Press, 1955.

Johnson, Thomas Carey. *Scientific Interests in the Old South.* New York: Appleton-Century, 1936.

Kaufman, Martin. *American Medical Education: The Formative Years, 1765–1910.* Westport, Conn.: Greenwood Press, 1976.

Kett, Joseph. *The Pursuit of Knowledge under Difficulties: From Self-Improvement to Adult Education in America, 1750–1990.* Stanford: Stanford University Press, 1994.

Kevles, Daniel J. *The Physicists: The History of a Scientific Community in Modern America.* Cambridge: Harvard University Press, 1995.

Khan, Bibi Zorina. "'The Progress of Science and the Useful Arts': Inventive Activity in the Antebellum Period." Ph.D. diss., UCLA, 1991.

Kirdahy, Carolyne. "The Morrill Land Grant Act and the Massachusetts Institute of Technology." MS, MIT, 1989.

Kohlstedt, Sally. "Parlors, Primers, and Public Schooling: Education for Science in Nineteenth Century America." *Isis* 81 (1990): 425–45.

———. "The Geologists' Model for National Science, 1840–1847." *Proceedings of the American Philosophical Society* 118 (1974): 179–95.

———. *The Formation of the American Scientific Community: The American Association for the Advancement of Science, 1846–1860.* Urbana: University of Illinois Press, 1976.

Kolesnik, Walter B. *Mental Discipline in Modern Education.* Madison: University of Wisconsin Press, 1958.

Kwik, Robert Julius. "The Function of Applied Science and the Mechanical Laboratory during the Period of Formation of the Profession of Mechanical Engineering, as Exemplified in the Career of Robert Henry Thurston, 1839–1903." Ph.D. diss., University of Pennsylvania, 1974.

Lane, Jack C. "The Yale Report of 1828 and Liberal Education: A Neorepublican Manifesto." *History of Education Quarterly* 27 (1987): 325–38.

Layton, Edwin. "Mirror-Image Twins: The Communities of Science and Technology in 19th Century America." *Technology and Culture* 12 (January 1971): 562–80.

Leslie, W. Bruce. *Gentlemen and Community: The College in the "Age of the University." 1865–1917* University Park: Pennsylvania State University Press, 1992.

Lessing, Peter. "The Rogers-Hotchkiss Geological maps of Virginia and West Virginia." *Earth Sciences History* 14 (1995): 84–97.

Levin, Miriam R. *Defining Women's Scientific Enterprise: Mount Holyoke Faculty and the Rise of American Science.* Hanover, N.H.: University Press of New England, 2005.

Livingstone, David N. *Nathaniel Southgate Shaler and the Culture of American Science.* Tuscaloosa: University of Alabama Press, 1987.

Lucas, Christopher J. *American Higher Education: A History.* New York: St. Martin's Press, 1994.

Lurie, Edward. "Louis Agassiz and the Idea of Evolution." *Victorian Studies* (September 1959): 87–108.

———. *Louis Agassiz: A Life in Science.* Baltimore: Johns Hopkins University Press, 1988.

———. *Nature and the American Mind: Louis Agassiz and the Culture of Science.* New York: Science History Publications, 1974.

Mack, Charles R., and Henry H. Lesesne, eds. *Francis Lieber and the Culture of Mind.* Columbia: University of South Carolina Press, 2005.

Malone, Dumas. *The Sage of Monticello.* Charlottesville: University of Virginia Press, 2005.

Manning, Thomas G. *Government in Science: The U.S. Geological Survey, 1867–1894.* Lexington: University of Kentucky Press, 1967.

Mannix, Loretta H. "Communications to the Society of Arts at its Regular Meetings Beginning with the Meeting of December 12, 1862." MS, MIT Archives, 1979.

Marsden, George M. *The Soul of the American University: From Protestant Establishment to Established Non-Belief.* New York: Oxford University Press, 1994.

Maslanka, John S. "William Barton Rogers' Conception of an Institute of Technology." B.S. thesis, MIT, 1961.

McCardell, John. *The Idea of a Southern Nation: Southern Nationalists and Southern Nationalism, 1830–1860.* New York: W. W. Norton, 1979.

McLachlan, James. "The American College in the Nineteenth Century: Toward a Reappraisal." *Teacher's College Record* 80 (1978): 287–306.

McPherson, James. *Battle Cry of Freedom: The Civil War Era.* New York: Oxford University Press, 1988.

McWhiney, Grady. *Cracker Culture: Celtic Ways in the Old South.* Tuscaloosa: University of Alabama Press, 1988.

Meier, Hugo. "Technology and Democracy, 1800–1860." *Mississippi Valley Historical Review* 43 (1957): 618–40.

Millbrooke, Anne Marie. "State Geological Surveys of the Nineteenth Century." Ph.D. diss, University of Pennsylvania, 1981.

Miller, Lilian B. *The Lazzaroni: Science and Scientists in Mid-Nineteenth Century America.* Washington, D.C.: Smithsonian Institution Press, 1972.

Miller-Bernal, Leslie. *Separate by Degree: Women Students' Experiences in Single-Sex and Coeducational Colleges.* New York: Peter Lang, 2000.

Molella, Arthur P., and Nathan Reingold. "Theorists and Ingenious Mechanics: Joseph Henry Defines Science." *Science Studies* 3 (October 1973): 323–51.

Montgomery, Scott L. *Minds for the Making: The Role of Science in American Education, 1750–1990.* New York: Guilford Press, 1994.

Morison, Samuel Eliot. *The Oxford History of the United States.* Oxford: Oxford University Press, 1927.

———. *Three Centuries of Harvard, 1636–1936.* Cambridge: Harvard University Press, 1936.

———, ed. *The Development of Harvard University since the Inauguration of President Eliot, 1869–1929.* Cambridge: Harvard University Press, 1930.

Morrison, James L., Jr. "Educating the Civil War Generals: West Point, 1833–1861." *Military Affairs* 38 (1974): 108–11.

Munroe, James P. *A Life of Francis Amasa Walker.* New York: Henry Holt, 1923.

———. *Discussions in Education by Francis A. Walker, Ph.D., LL.D.* New York: Henry Holt, 1899.

Murphy, Lamar Riley. *Enter the Physician: The Transformation of Domestic Medicine, 1760–1860.* Tuscaloosa: University of Alabama Press, 1991.

Nartonis, David K. "Louis Agassiz and the Platonist Story of Creation at Harvard, 1795–1846." *Journal of the History of Ideas* 66 (2005): 437–49.

National Academy of Sciences. *A History of the First Half-Century of the National Academy of Sciences, 1863–1913.* Washington, D.C.: National Academy of Sciences, 1913.

Nemec, Mark R. *Ivory Towers and Nationalist Minds: Universities, Leadership, and the Development of the American State.* Ann Arbor: University of Michigan Press, 2006.

Noble, David. *America by Design: Science, Technology, and the Rise of Corporate Capitalism.* New York: Knopf, 1977.

Noel, Lisa. *Intolerance: A General Survey.* Montreal: McGill University Press, 1994.

Nolan, John Patrick. "Genteel Attitudes in the Formation of the American Scientific Community: The Career of Benjamin Silliman of Yale." Ph.D. diss, Columbia University, 1978.

Noll, Mark A. *Princeton and the Republic, 1768–1822: The Search for a Christian Enlightenment in the Era of Samuel Stanhope Smith.* Princeton: Princeton University Press, 1989.

Novak, Steven J. *The Rights of Youth: American Colleges and Student Revolt, 1789–1815.* Cambridge: Harvard University Press, 1977.

Numbers, Ronald. *Darwinism Comes to America.* Cambridge: Harvard University Press, 1998.

————, ed. *The Education of American Physicians.* Berkeley: University of California Press, 1980.

Numbers, Ronald L., and Janet S. Numbers. "Science in the Old South: A Reappraisal." *Journal of Southern History* 48 (1982): 163–84.

Numbers, Ronald L., and John Stenhouse, eds. *Disseminating Darwinism: The Role of Place, Race, Religion, and Gender.* New York: Cambridge University Press, 2000.

Nye, Mary Jo. *Before Big Science: Pursuit of Modern Chemistry and Physics.* Cambridge: Harvard University Press, 1999.

O'Brien, Michael. *Conjectures of Order: Intellectual Life and the American South, 1810–1860.* 2 vols. Chapel Hill: University of North Carolina Press, 2004.

O'Connor, Thomas H. *Civil War Boston: Home Front and Battle Field.* Boston: Northeastern University Press, 1997.

Oleson, Alexandra, and Sanborn C. Brown, eds. *The Pursuit of Knowledge in the Early American Republic: American Scientific and Learned Societies from Colonial Times to the Civil War.* Baltimore: Johns Hopkins University Press, 1976.

Osborne, Ruby Orders. *The Crisis Years: The College of William and Mary in Virginia, 1800–1827.* Richmond, Va.: Deitz Press, 1989.

Pace, Robert F., and Christopher A. Bjornsen. "Adolescent Honor and College Student Behavior in the Old South." *Southern Cultures* (Fall 2000): 9–28.

————. *Halls of Honor: College Men in the Old South.* Baton Rouge: Louisiana State University Press, 2004.

Parshall, Karen Hunger. *James Joseph Sylvester: Jewish Mathematician in a Victorian World.* Baltimore: Johns Hopkins Press, 2006.

————. *James Joseph Sylvester: Life and Works in Letters.* New York: Oxford University Press, 1998.

Parshall, Karen Hunger, and David E. Rowe. *The Emergence of the American Mathematical Research Community, 1878–1900: J. J. Sylvester, Felix Klein, and E. H. Moore.* Providence, R.I.: American Mathematical Society, 1994.

Potts, David. "American College in the Nineteenth Century: From Localism to Denominationalism." *History of Education Quarterly* 11 (1971): 363–80.

————. "Curriculum and Enrollments: Some Thoughts Assessing the Popularity of Antebellum Colleges." *History of Higher Education Annual* 1 (1981): 88–109.

Prescott, Samuel C. *When MIT Was "Boston Tech," 1861–1916.* Cambridge, Mass.: MIT Press, 1954.

Pursell, Caroll W., Jr., ed. *Technology in America: A History of Individuals and Ideas.* Cambridge, Mass.: MIT Press, 1981.

————. *Readings in Technology and American Life.* New York: Oxford University Press, 1969.

Pyenson, Lewis. "An End to National Science: The Meaning and the Extension of Local Knowledge." *History of Science* (September 2002): 272–90.

Reingold, Nathan, ed. *Science in America since 1820.* New York: Science History Publications, 1976.

Reuben, Julie. *The Making of the Modern University: Intellectual Transformation and the Marginalization of Morality.* Chicago: University of Chicago Press, 1996.

Reynolds, Terry S. "The Education of Engineers in America before the Morrill Act of 1862." *History of Education Quarterly* 32 (1992): 459–82.

Rezneck, Samuel. *Education for a Technological Society: A Sesquicentennial History of Rensselaer Polytechnic Institute.* Troy, N.Y.: RPI Press, 1968.

Ricketts, Palmer C. *History of the Rensselaer Polytechnic Institute, 1824–1894.* New York: Wiley, 1895.

———. *Rensselaer Polytechnic Institute.* Troy, N.Y.: RPI Press, 1933.

———. *Rensselaer Polytechnic Institute: A Short History.* Troy, N.Y.: RPI Press, 1930.

Roberts, Jon H., and James Turner. *The Sacred and the Secular University.* Princeton: Princeton University Press, 2000.

Roller, Duane H. D., ed. *Perspective in the History of Science and Technology.* Norman: University of Oklahoma Press, 1971.

Roper, Peter W. "Jed Hotchkiss and the Geological Map of Virginia." *Earth Sciences History* 10 (1991): 38–43.

Ross, Dorothy. *The Origins of American Social Science.* New York: Cambridge University Press, 1991.

Rossiter, Margaret W. *Women Scientists in America: Struggles and Strategies to 1940.* Baltimore: Johns Hopkins University Press, 1982.

———. "Louis Agassiz and the Lawrence Scientific School." B.S. thesis, Harvard University, 1966.

Rothblatt, Sheldon, and Bjorn Wittrock, eds. *The European and American University since 1800: Historical and Sociological Essays.* New York: Cambridge University Press, 1993.

Rothstein, William G. *American Medical Schools and the Practice of Medicine: A History.* New York: Oxford University Press, 1987.

Rouse, Parker, Jr. *Virginia: The English Heritage in America.* New York: Hastings House, 1976.

Rudolph, Frederick. *Curriculum: A History of the American Undergraduate Course of Study since 1636.* San Francisco: Jossey-Bass, 1977.

———. *The American College and University: A History.* New York: Knopf, 1962.

Ruegg, Walter. *Universities in the Nineteenth and Early Twentieth Centuries, 1800–1945.* New York: Cambridge University Press, 2004.

Schurter, William John. "The Development of the Russian System of Tool Instruction (1763–1893) and Its Introduction into U.S. Industrial Education Programs (1876–1893)." Ph.D. diss., University of Maryland, 1982.

Segal, Howard P. *Technological Utopianism in American Culture.* Chicago: University of Chicago Press, 1985.

Shade, William. *Democratizing the Old Dominion: Virginia and the Second Party System, 1824–1861.* Charlottesville: University Press of Virginia, 1996.

Sheets-Pyenson, Susan. *John William Dawson: Faith, Hope, and Science.* Montreal: McGill-Queen's University Press, 1996.

Shoemaker, Philip Stanley. "Stellar Impact: Ormsby Macknight Mitchel and Astronomy in Antebellum America." Ph.D. diss., University of Wisconsin, Madison, 1991.

Shrock, Richard Rakes. *Geology at MIT, 1865–1965: A History of the First Hundred Years of Geology at Massachusetts Institute of Technology.* 2 vols. Cambridge, Mass.: MIT Press, 1977–82.

Silliman, Robert H. "The Richmond Boulder Trains: Verae Causae in 19th-Century American Geology." *Earth Sciences History* 10 (1991): 60–72.

Silverman, Robert, and Mark Beach, "A National University for Upstate New York." *American Quarterly* 22 (1970): 701–13.

Sinclair, Bruce. *Philadelphia's Philosopher Mechanics: A History of the Franklin Institute, 1824–1865.* Baltimore: Johns Hopkins University Press, 1974.

———, ed. *New Perspectives on Technology and American Culture.* Philadelphia: American Philosophical Society, 1986.

Slaughter, Sarah. "The Origins of MIT." MS, MIT, 1980.

Slotten, Hugh R. "Science, Education, and Antebellum Reform: The Case of Alexander Dallas Bache." *History of Education Quarterly* 31 (1991): 323–42.

———. *Patronage, Practice, and the Culture of American Science: Alexander Dallas Bache and the U.S. Coast Survey.* New York: Cambridge University Press, 1994.

Smith, Alfred Glaze. *Economic Readjustment of an Old Cotton State: South Carolina, 1820–1860.* Columbia: University of South Carolina Press, 1958.

Smith, Harold. *The Society for the Diffusion of Useful Knowledge 1826–1846: A Social and Bibliographical Evaluation.* London: Vine Press, 1974.

Solomon, Barbara Miller. *In the Company of Educated Women: A History of Women and Higher Education in America.* New Haven: Yale University Press, 1985.

Stephens, Lester D. *Joseph LeConte: Gentle Prophet of Evolution.* Baton Rouge: Louisiana State University Press, 1982.

———. *Science, Race, and Religion in the American South: John Bachman and the Charleston Circle of Naturalists, 1815–1895.* Chapel Hill: University of North Carolina Press, 2000.

Stevens, Edward W., Jr. *The Grammar of the Machine: Technical Literacy and Early Industrial Expansion in the United States.* New Haven: Yale University Press, 1995.

Stevenson, Louise L. *Scholarly Means to Evangelical Ends: The New Haven Scholars and the Transformation of Higher Learning in America, 1830–1890.* Baltimore: Johns Hopkins University Press, 1986.

Storr, Richard J. *The Beginnings of Graduate Education in America.* New York: Arno Press, 1969.

Story, Ronald. *The Forging of an Aristocracy: Harvard and the Boston Upper Class, 1800–1870.* Middletown, Conn.: Wesleyan University Press, 1980.

Stratton, Julius A., and Loretta Mannix. *Mind and Hand: The Birth of MIT.* Cambridge, Mass.: MIT Press, 2005.

Tachikawa, Akira. "The Two Sciences and Religion in Antebellum New England: The Founding of the Museum of Comparative Zoology and the Massachusetts Institute of Technology." Ph.D. diss., University of Wisconsin, Madison, 1978.

Taylor, Natalie A. "The Ante-Bellum College Movement: A Reappraisal of Tewksbury's Founding of American Colleges and Universities." *History of Education Quarterly* 13 (1973): 261–74.

Thelin, John R. *A History of American Higher Education.* Baltimore: Johns Hopkins University Press, 2004.

Thorton, Tamara Plakins. *Cultivating Gentlemen: The Meaning of Country Life among the Boston Elite, 1785–1860.* New Haven: Yale University Press, 1989.

Tise, Larry. *Proslavery: A History of the Defense of Slavery in America, 1701–1840.* Athens: University of Georgia Press, 1987.

Tolley, Kim. "Science for Ladies, Classics for Gentlemen: A Comparative Analysis of Scientific Subjects in the Curricula of Boys' and Girls' Secondary Schools in the United States, 1794–1850." *History of Education Quarterly* 36 (1996): 129–53.

———. *The Science Education of American Girls: A Historical Perspective.* New York: Routledge-Falmer, 2003.

Trecker, Janice Law. "Sex, Science and Education." *American Quarterly* 26 (1974): 352–66.

Turner, Paul Venable. *Campus: An American Planning Tradition.* Cambridge, Mass.: MIT Press, 1984.

Tyack, David. *George Ticknor and the Boston Brahmins.* Cambridge: Harvard University Press, 1967.

Tyack, David, and Elizabeth Hansot. *Learning Together: A History of Coeducation in American Public Schools.* New York: Russell Sage Foundation, 1992.

Veysey, Lawrence. *The Emergence of the American University.* Chicago: University of Chicago Press, 1965.

Wagoner, Jennings L. "Honor and Dishonor at Mr. Jefferson's University: The Antebellum Years." *History of Education Quarterly* 26 (Summer 1986): 155–79.

———. *Jefferson and Education.* Charlottesville, Va.: Thomas Jefferson Foundation, 2004.

Wall, Charles Coleman, Jr. "Students and the Student Life at the University of Virginia, 1825 to 1861." Ph.D. diss., University of Virginia, 1979.

Walsh, Richard, and William Lloyd Fox. *Maryland: A History, 1632–1974.* Baltimore: Maryland Historical Society, 1974.

Warner, Deborah Jean. "Science Education for Women in Antebellum America." *Isis* 69 (1978): 58–67.

Weiss, John Hubbel. *The Making of Technological Man: The Social Origins of French Engineering Education.* Cambridge, Mass.: MIT Press, 1982.

Wiley, Wayne Hamilton. "Academic Freedom at the University of Virginia: The First Hundred Years—From Jefferson through Alderman." Ph.D. diss., University of Virginia, 1973.

Williams, Roger L. *The Origins of Federal Support for Higher Education: George W. Atherton and the Land-Grant College Movement.* University Park: Pennsylvania State University Press, 1991.

Wilson, Leonard G. "The Emergence of Geology as a Science in the United States." *Journal of World History* 10 (1967): 416–37.

Wilson, Leonard G. *Lyell in America: Transatlantic Geology, 1841–1853.* Baltimore: Johns Hopkins University Press, 1998.

———, ed. *Benjamin Silliman and His Circle: Studies on the Influence of Benjamin Silliman* on Science in America: Prepared in Honor of Elizabeth H. Thomson.* New York: Science History Publications, 1979.

Winsor, Mary P. "Agassiz's Notions of a Museum: The Vision and the Myth." In Michael T. Ghiselin and Alan E. Leviton, eds. *Cultures and Institutions of Natural History: Essays in the History and Philosophy of Science.* San Francisco: California Academy of Sciences, 2000. 249–71.

————. *Reading the Shape of Nature: Comparative Zoology at the Agassiz Museum* Chicago: University of Chicago Press, 1991.

Winterer, Caroline. *The Culture of Classicism: Ancient Greece and Rome in American Intellectual Life, 1780–1910.* Baltimore: Johns Hopkins University Press, 2002.

Wyatt-Brown, Bertram. *Honor and Violence in the Old South.* New York: Oxford University Press, 1986.

Yates, R. C. "Sylvester at the University of Virginia." *American Mathematical Monthly* 44 (1937): 194–201.

Zschoche, Sue. "Dr. Clarke Revisited: Science, True Womanhood, and Female Collegiate Education." *History of Education Quarterly* 29 (1989): 545–69.